T0253997

# EVOLUTION AND RATIONALITY

This volume explores from multiple perspectives the subtle and interesting relationship between the theory of rational choice and Darwinian evolution. In rational choice theory, agents are assumed to make choices that maximize their utility; in evolution, natural selection 'chooses' between phenotypes according to the criterion of fitness maximization. So there is a parallel between utility in rational choice theory, and fitness in Darwinian theory. This conceptual link between fitness and utility is mirrored by the interesting parallels between formal models of evolution and rational choice. The essays in this volume, by leading philosophers, economists, biologists and psychologists, explore the connection between evolution and rational choice in a number of different contexts, including choice under uncertainty, strategic decision making and prosocial behaviour. They will be of interest to students and researchers in philosophy of science, evolutionary biology, economics and psychology.

SAMIR OKASHA is Professor of Philosophy of Science at the University of Bristol. He is the author of *Philosophy of Science: A Very Short Introduction* (2002) and *Evolution and the Levels of Selection* (2006).

KEN BINMORE is Professor Emeritus of Economics at University College, London, and a Visiting Emeritus Professor of Economics at the University of Bristol. He is the author of *Natural Justice* (2005), *Game Theory: A Very Short Introduction* (2007), and *Rational Decisions* (2008).

# EVOLUTION AND RATIONALITY

*Decisions, Co-operation and Strategic Behaviour*

EDITED BY

SAMIR OKASHA AND KEN BINMORE

CAMBRIDGE
UNIVERSITY PRESS

# CAMBRIDGE
## UNIVERSITY PRESS

University Printing House, Cambridge CB2 8BS, United Kingdom

Published in the United States of America by Cambridge University Press, New York

Cambridge University Press is part of the University of Cambridge.

It furthers the University's mission by disseminating knowledge in the pursuit of
education, learning and research at the highest international levels of excellence.

www.cambridge.org
Information on this title: www.cambridge.org/9781107416840

© Cambridge University Press 2012

First published 2012
First paperback edition 2014

*A catalogue record for this publication is available from the British Library*

ISBN 978-1-107-00499-3 Hardback
ISBN 978-1-107-41684-0 Paperback

# Contents

# *Figures and Tables*

## FIGURES

## TABLES

# Contributors

SIEGFRIED BERNINGHAUS is a Professor in the Department of Economics, Karlsruhe Institute of Technology.

HENRY BRIGHTON is a Research Scientist at the Center for Adaptive Behavior and Cognition, Max Planck Institute for Human Development, Berlin.

MAXWELL BURTON-CHELLEW is a Postdoctoral Researcher, Department of Zoology, University of Oxford.

CLAIRE EL MOUDEN is a Postdoctoral Prize Research Fellow at Nuffield College and Research Fellow, Department of Zoology, University of Oxford.

ANDY GARDNER is a Royal Society University Research Fellow, Department of Zoology, University of Oxford.

GERD GIGERENZER is Director of the Center for Adaptive Behavior and Cognition, Max Planck Institute for Human Development, Berlin.

HERBERT GINTIS is Professor of Economics, Central European University, and at the Santa Fe Institute.

NATALIE GOLD is a Senior Research Fellow at King's College, London.

WERNER GÜTH is Director of the Strategic Interaction Group, Max Planck Institute of Economics, Jena.

PETER HAMMERSTEIN is Professor at the Institute for Theoretical Biology, Humboldt University, Berlin.

ALASDAIR I. HOUSTON is Professor of Theoretical Biology at the University of Bristol.

SIMON M. HUTTEGGER is Associate Professor of Philosophy, University of California, Irvine.

JULIAN JAMISON is Senior Economist at the Center for Behavioral Economics, Federal Reserve Bank of Boston.

HARTMUT KLIEMT is Professor of Philosophy and Economics, Frankfurt School of Finance and Management.

KIM STERELNY is Professor of Philosophy at the Australian National University.

JACK VROMEN is Professor of Theoretical Philosophy at Erasmus University, Rotterdam.

STUART A. WEST is Professor of Evolutionary Biology, Department of Zoology, University of Oxford.

DAVID H. WOLPERT is at the Santa Fe Institute.

KEVIN J. S. ZOLLMAN is Assistant Professor of Philosophy, Carnegie Mellon University.

# Introduction

## Samir Okasha and Ken Binmore

There exist deep and interesting connections, both thematic and formal, between evolutionary theory and the theory of rational choice, despite their apparently different subject matters. These connections arise because a notion of optimization or maximization is central to both areas. In rational choice theory, agents are assumed to make choices that maximize their utility, while in evolutionary theory, natural selection 'chooses' between alternative phenotypes, or genes, according to the criterion of fitness maximization. As a result, evolved organisms often exhibit behavioural choices that appear designed to maximize their fitness, which suggests that the principles of rational choice might be applicable to them. This conceptual link between evolution and rational choice explains the fascinating exchange of ideas between evolutionary biology and economics that has taken place in the last forty years, particularly in relation to decision making under uncertainty, and strategic interaction.

The chapters in this book all deal with aspects of the evolution/rationality relationship, from a range of perspectives. The book emerged from a series of workshops and conferences held at the University of Bristol between 2008 and 2011, under the auspices of the 'Evolution, Cooperation and Rationality' research project, funded by the Arts and Humanities Research Council of the UK and directed by ourselves. The project examined foundational and conceptual issues arising from recent work on social behaviour, decision making and strategic interaction, and had a strongly interdisciplinary orientation. This is reflected in the composition of the book – the authors include leading researchers in evolutionary biology, philosophy of science, experimental economics, game theory and psychology. The result illustrates the rich diversity of approaches to the study of evolution and rationality, and, we hope, will help promote constructive dialogue between them.

The fundamental paradigm of the economic theory of rational choice is that someone who chooses 'consistently', where this means conforming to certain rather intuitive axioms, behaves as though maximizing expected utility. Biologists similarly argue that in suitably idealized circumstances, evolution will produce animals that behave as though maximizing their expected fitness, where 'fitness' refers to the additional number of off-spring the animal produces as a result of the behaviour in question. So the question arises: when it is possible to identify the economists' notion of utility with the biologists' notion of fitness?

This question of the relationship between utility and fitness is a central theme in a number of chapters here, and is in the background of most of the others. Kim Sterelny defends the idea that fitness maximiza-tion and utility maximization can sometimes be expected to coincide in human populations, but not always. It depends on whether information is transmitted vertically or horizontally, and on whether group selection is or is not a prevalent factor, Sterelny argues, for these determine whether population-level processes will be effective in binding proximal motiv-ation to fitness consequences. The utility versus fitness issue is also cen-tral to the chapter by Claire El Mouden, Maxwell Burton-Chellew, Andy Gardner and Stuart A. West, who ask what quantity – if any – we should expect humans to appear designed to maximize. Their analysis is based on inclusive fitness theory, the highly successful approach to social behav-iour devised by W. D. Hamilton. El Mouden et al. make the case that humans, like other social animals, have been selected to maximize their inclusive fitness. They admit that much human behaviour doesn't appear to actually achieve this, but consider a number of reasons, quite different from Sterelny's, for how to square the data with the assumption of fitness maximization.

Alasdair Houston's chapter also tackles the issue of fitness and utility, from a very different perspective. A key assumption in rational choice theory is that an agent's preferences or choices should be transitive; other-wise, the agent cannot be represented as a utility maximizer. However, systematic intransitivities of choice have been reported in both human and non-human subjects; this raises the question of how such apparently irrational behaviour could have evolved. Houston examines a number of potential explanations for how natural selection can result in intransi-tive choices. Some of these explanations, though not all of them, imply that the choice behaviour is only apparently irrational – in that transi-tivity can be restored so long as the modeller takes a 'correct view' of the decision maker's options. In particular, Houston stresses the importance of considering 'state-dependent' decisions, in which an animal's choice behaviour is partly determined by a state variable, e.g. its energy reserves.

If we neglect state dependence, we may be led to see an animal's choice behaviour in any one state as irrational, when in fact it is implementing an optimal state-dependent strategy.

The idea that there is a fairly straightforward evolutionary basis for rational choice maxims such as consistency of preferences and maximization of expected utility is developed in this volume by Herbert Gintis, in his chapter on the unification of the behavioural sciences from an evolutionary perspective. Gintis defends the 'rational actor' model that underpins most economic theory, and argues that it integrates seamlessly with evolutionary biology. He rejects the view, held by many psychologists, that humans exhibit systemic cognitive biases and irrationalities which undermine the applicability of rational choice theory. General evolutionary considerations tell against this pessimistic idea, Gintis argues, and the experimental data can be explained in ways that are compatible with the rational actor model. However, a complete theory of behavioural choice must go beyond the rational actor model, he thinks, and incorporate ideas from both evolutionary biology and social psychology. Gintis outlines how he thinks this conceptual unification should take place.

A quite different attitude towards rational choice theory is found in Henry Brighton and Gerd Gigerenzer's chapter, who focus on a problem faced by all organisms: making inductive inferences in an uncertain world. Their chapter builds on Gigerenzer's previous work in which he strongly criticizes the use of optimality and rational choice models to understand adaptive behaviour, arguing that 'simple heuristics' will often outperform attempts at maximization. Brighton and Gigerenzer defend this view in relation to inductive inference. They argue that rational choice models only work in what L. J. Savage called 'small worlds', i.e. situations where the state space of the decision problem is pregiven, but are highly misleading in 'large worlds', where the state space must itself be inferred from observations. In large worlds, adaptive behaviour is easier to produce via heuristics than optimization. Gigerenzer and Brighton connect this argument with an interesting philosophical claim, namely that there is no such thing as a 'one true rationality', since rationality principles are invented, not discovered. They suggest an understanding of 'rational' and 'optimal' which is compatible with this philosophy.

An area where the interplay of ideas between economics and evolution has been particularly fruitful is game theory. Originally designed to explain the strategic choices of rational human agents, game theory was introduced into biology in the 1970s and 1980s to explain aspects of animal behaviour. The basic concept of traditional game theory is the idea of a Nash equilibrium. A profile of strategies – one for each player – is a Nash equilibrium if no player has an incentive to deviate from his strategy provided that nobody else deviates first. In traditional game theory, the Nash equilibrium is interpreted as an equilibrium in rational deliberation, i.e. a situation from which no rational player will unilaterally deviate. Thus we can predict that rational players in a game will end up at a Nash equilibrium of that game. However, the Nash equilibrium also admits of an evolutionary interpretation, because any dynamical process that always moves in the direction of higher payoffs can only stop when it gets to a Nash equilibrium.

This dual interpretation has proved very useful in evolutionary biology, because it sometimes allows theorists to use the rational interpretation to predict the outcome of an evolutionary process without needing to study the complicated details of the process itself. When reasoning in this way, biologists usually speak of an *evolutionarily stable strategy* (ESS), a concept first devised by John Maynard Smith and George Price and which bears a close relation to the Nash equilibrium concept. An ESS refers to a state of a population that, once reached, cannot be invaded by small groups of mutants.

Peter Hammerstein's chapter traces the fascinating intellectual history of how game theory entered evolutionary biology, with a focus on conceptual and foundational issues. He makes a strong case for the power of strategic analysis in biology, citing numerous examples of biological phenomena that have been illuminated through the application of game-theoretic methods, including conflicts over parental investment and intraorganismic conflict. Despite these success stories, and despite the fact that biological game theory can dispense with the improbably strong epistemic assumptions made by classical game theory – such as common knowledge of rationality – Hammerstein sounds a note of caution. Biologists cannot simple appeal to the 'authority of traditional game theory' to analyse the consequences of strategic interaction, he argues; explicit attention to the evolutionary dynamics, rather than merely looking for stable equilibria, may be required.

This issue of evolutionary dynamics, and the relation between rational and evolutionary game theory, are also central to the chapter by Simon M. Huttegger and Kevin J. S. Zollman. They offer a searching critique of what they call 'ESS methodology' in biology, i.e. the practice of assuming that evolution will take a population to an ESS, and using this to guide the interpretation of observed biological phenomena. The problem with this methodology is that in some circumstances, natural selection will not carry a population to an ESS state, so the methodology is at best fallible; there is no short cut to studying the full evolutionary dynamics, they argue. Huttegger and Zollman offer a striking illustration of this point with a simple 'sender–receiver' signalling game, in which most initial population states converge to an equilibrium that is not an ESS. Interestingly, Huttegger and Zollman trace the failure of the ESS methodology to a feature that the ESS concept shares with other 'refinements' (or logical strengthenings) of the Nash equilibrium concept discussed in traditional rational game theory.

Many authors have contrasted rational and evolutionary game theory, or viewed them as rival 'interpretations' of an underlying formalism. An alternative approach is to try to integrate the two. This is the route taken by Siegfried Berninghaus, Werner Güth and Hartmut Kliemt, who advocate what they call an 'indirect evolutionary approach'. (Somewhat similar ideas have also been developed under the heading 'evolution of preferences'.) The core of Berninghaus et al.'s indirect evolutionary approach is to model strategic behaviour using two timescales, long and short, corresponding to ultimate and proximate levels of explanation. In the short term, agent's subjective preferences, and thus their behavioural choices, may be governed by rules and social norms, rather than by the pursuit of Darwinian fitness. But over a longer evolutionary timescale, the proliferation of different behaviours depends on their 'objective payoffs', or fitnesses. Berninghaus et al. show that subjective motives other than maximization of objective payoff can be favoured (a result that tallies with Sterelny's argument). The indirect evolutionary approach offers an interesting take on the relationship between agents' subjective motivations and the evolutionary consequences of their actions.

The 'two-timescales' idea also features in the chapter by David H. Wolpert and Julian Jamison, on the strategic choices of 'non-rational' players. In their version, however, the longer timescale corresponds to learning within the lifetime of a single player, rather than an evolutionary process unfolding over multiple generations. (It is well known that evolution and learning exhibit interesting parallels, a point discussed by Hammerstein.) Focusing on learning rather than evolution permits Wolpert and Jamison to stick with the Nash equilibrium concept, rather than the ESS concept which is harder to work with. Wolpert and Jamison's central idea is that of a 'persona game', in which a player chooses a persona, e.g. that of someone who refuses to be treated unfairly, signals the persona to others, and commits to using it during the play of a game. (This need not be done fully consciously.) Wolpert and Jamison argue that humans have a remarkable ability to adopt different personae in their social interactions, and explore the subtle strategic implications of this ability. They show how various forms of 'non-rational' behaviour, such as co-operation in a one-shot prisoner's dilemma, can be explained via persona games, and argue that this explanation fits the extant data.

Co-operative behaviour, and the problem of how to incorporate it into a systematic framework, is also central in Natalie Gold's chapter. She explores the notion of 'team reasoning', originally due to Michael Bacharach and Bob Sugden, as a potential explanation of apparently non-rational choices in games such as the prisoner's dilemma. In standard game theory, the players reason in an 'individualistic' way, aiming to maximize their own utility (though of course their utility function may be 'other regarding'). In team reasoning, players are able to mentally identify with a particular team, or set of players, and make choices that are optimal from the point of view of the whole team. Gold explores two subtly different versions of team reasoning, and shows how it can lead players to co-operate in social dilemmas. It is particularly interesting that the basic idea behind team reasoning – invoking 'team payoff' in addition to individual payoff – can also be found in evolutionary biology, in the theory of multilevel selection, a point that Gold discusses.

Co-operation and social behaviour are also central to the chapter by Jack Vromen, which is a philosophical investigation of recent work on 'strong reciprocity' in humans. Strong reciprocity refers to our predisposition to co-operate with others, and to punish others who fail to co-operate, even when this punishment is costly to administer. There is considerable evidence, both experimental and anthropological, in favour of the idea that humans are strong reciprocators in this sense. Vromen argues that the literature on strong reciprocity contains a number of conceptual confusions, in particular over whether it constitutes altruism or selfishness, and whether it requires group-level selection to evolve. He traces these confusions to a failure to keep separate the evolutionary and the psychological meanings of 'altruism', an issue closely connected to the distinction between proximate and ultimate explanation. He also examines the psychological evidence on our propensity to engage in costly punishment, arguing that it cannot resolve the issue of whether this propensity is psychologically selfish or altruistic.

This work was supported by the Arts and Humanities Research Council, grant no. AH/FO17502/1, and the European Research Council Seventh Framework Program (FP7/2007–2013), ERC Grant agreement no. 295449.

# CHAPTER I

# Towards a Darwinian theory of decision making
## Games and the biological roots of behavior

### Peter Hammerstein

## I.I INTRODUCTION

Conventional decision theory is normative and it attempts to identify decisions that are in some sense optimal. The decision maker is often assumed to have all the mental capabilities that real human beings can only dream of. Classical game theory has built on this approach and many of its scholars have almost routinely referred to the normative character of their theory as an excuse for the lack of empirical content. I claim that this excuse is unconvincing. Even an entirely rational visitor from outer space would meet real people on earth and would have to deal with them in a smart way. This visitor would be forced to learn as much as possible about the evolved psychology of humans in order to identify his best decisions in our world of the not-so-smart. It is therefore impossible to separate normative and descriptive approaches unless game theory deals exclusively with rational visitors from outer space.

In this chapter, I wish to explain how game theory can be firmly rooted in the life sciences without dismissing the legacy of its founders. I first take a look at the history of ideas in game theory and give my comments as a theoretical biologist. The next step is to explain the interesting links between reasoning in decision theory and those properties of the evolutionary process that look to us *as if* evolution itself were able to reason about decision problems. A subsequent excursion into the bacterial world demonstrates that even microbes reflect this feature of evolution. Looking finally at animal interactions, I discuss how basic ideas in game theory sometimes hit the nail on the head in relation to empirical findings and are at other times very misleading. A concluding remark is devoted to learning and the future of game theory.

I am grateful to Benjamin Bossan and Arnulf Koehncke for numerous comments I received when preparing this chapter.

## 1.2  A BIOLOGIST'S LOOK AT THE STRUGGLE FOR
## CONCEPTS IN CLASSICAL GAME THEORY

As a scientific discipline game theory emerged early in the twentieth century and gained general visibility in 1944 when John von Neumann and Oskar Morgenstern published their seminal book *Theory of Games and Economic Behavior*. The new discipline was meant to provide a framework for mathematical modeling in economics and the social sciences. Its development was, therefore, driven by the need to capture the essentials of decision making in interactive situations. Obviously, this need could not be satisfied by simply borrowing ideas from physics or any of the other natural sciences. Conversely, game theory later became the first formalized field of the social sciences that had considerable impact on theory development in a natural science. This was the case when evolutionary game theory emerged in biology (Maynard Smith and Price 1973; Maynard Smith 1982; Hammerstein and Selten 1994).

A game is, technically speaking, a mathematical model of an interaction with two or more actors (players) involved. Von Neumann and Morgenstern introduced basic forms of models that are still in use, such as the extensive and normal (strategic) form of a game. The founders of game theory were less successful, though, in foreshadowing the solution concepts (ways of analyzing games) that later became mainstream practice. Why did they not anticipate more of the theory for which they had paved the road? Undoubtedly, John von Neumann was a person with great visionary power and exactly for this reason he probably appreciated the difficulty of the following fundamental question: What can be regarded as a player's adequate strategic response to the strategy choices of other players if – as is usual – those choices are *unknown*? (Note here that a strategy is generally far more than a simple observable act, and even the choice of a simple act can only be observed after it has taken place.)

In my view, it is this *unknown* that should have puzzled other game theoreticians more than it typically did. Von Neumann demonstrated his appreciation of the strategic response problem by introducing a solution concept that avoids speculating about the unknown. Unfortunately, this avoidance led him to a very pessimistic decision principle, which is to "minimize maximum losses" by playing so-called *minimax strategies*. His minimax solution concept seems unconvincing from a modern point of view except for the context of two-person zero-sum games. The main criticism is that "minimaxers" ignore how likely it is that the persons they interact with will choose different behavioral options. Here, a defender of minimax could reply that *uncertainty* exists with regard to these probabilities. The defender would perhaps claim that one should take into account only those probabilities that are known, like that of rolling a given number with a single throw of dice. Only in this case of known probabilities is the risk calculable. This attempt to steer the conceptual discourse is debatable, however, since it assumes a clear distinction between decision under risk and uncertainty. At least from a Bayesian point of view such a distinction cannot be made, and rationality axioms in the footsteps of Savage (1954) would force a decision maker to form subjective probabilities in every decision situation.

The beauty of Bayesian decision theory should not prevent us from realizing that in most real-life situations it seems technically impossible for humans to practice with their evolved brains what Bayesian axioms would require them to do. Selten (1991) therefore expressed the view that Bayesian theory can neither describe typical human decision making, nor can it frequently be of practical normative use. He only admits that the application of Bayesian methods can make sense in special contexts. In the information processing of an insurance company, for example, subjective probabilities may be generated reasonably well on the basis of actuarial tables.

As a scientist trying to capture reality I can hardly disagree with Selten's radical view on Bayesian concepts but also have to admit – as does Selten – that one can, if one wishes, think about fictitious rational beings that possess all the technical expertise and capacities needed for Bayesian decision making. As a thought experiment it may then be feasible to explore the interactions that could take place in this fictional world. But caution is more than strongly advised when returning to the world of facts.

Addressing now the most successful solution concept of classical game theory, the Nash equilibrium (Nash 1951), we run into the same problems as before. A Nash equilibrium specifies strategies for each player in such a way that no player could improve his expected payoff by unilaterally deviating from this specified strategy profile. Nash's concept looks rather trivial at first glance, when one realizes that it merely transfers the idea of optimization (here maximization of payoff) to interactive situations (every player responds optimally to the other players), in a simple and intuitive way. This may well be the reason why Nash as a mathematician considered it an obvious choice. His concept raises many questions, however, if one starts thinking about its deeper justification.

What assumptions are needed to back up the Nash equilibrium? This depends strongly on the perspective taken. Let us start by asking how a player could find a Nash equilibrium through some kind of careful reasoning rather than by intuition, routine learning, or teaching. In order to anticipate the actions of others, a carefully reasoning player would have to theorize about their minds, and – unfortunately for empirically void, idealized approaches – the minds of others may not be operating through careful reasoning. A sensible way of conducting one's thought would, therefore, be to rely on empirical knowledge about the evolved psychology of decision making. From behavioral experiments we know that this evolved psychology often does not favor Nash equilibria. When interacting with real people, rational players would thus frequently be forced to avoid Nash equilibria in order to play best responses to the strategies that actually matter, and that they actually encounter. Looked at from this angle, the Nash equilibrium fails to be convincing even as a normative concept: *we often ought not to do what classical game theory claims we ought to do.*

Now, in order to come up with a fairly general, reasoning-based justification of the Nash equilibrium, one has to invoke something like the following grandiose assumption:

(A1) All players in a game are rational and all know that all know that all are rational.

Why does this assumption help? Since all players are now artificial beings, void of psychology, and slaves of the axioms of rationality, they have no problem figuring out how everybody else will make strategic choices. "Reading the mind" can here simply be replaced by "reading the axioms." Consequently, if these rational players adopt a solution of how to play the game, this solution must be consistent with the axioms that define their fictitious minds. No solution can then be adopted that includes payoff incentives for deviation. Along this line of reasoning we can interpret the Nash equilibrium property as a *necessary condition for what qualifies as a rational decision in a purely rational world.*

Note that in special cases the Nash equilibrium is justified under assumptions weaker than (A1). For example, in games with strictly dominant strategies, such as the prisoner's dilemma, an optimal strategic decision can be made without assuming anything about the rationality or the knowledge of others. Note also that a group of players educated in classical game theory, all knowing about their joint education and trusting the success of "brainwashing" by their teachers, may have reasons to play a Nash equilibrium *because* of this education. In this sense, the central solution concept of classical game theory has a potential to become a self-fulfilling prophecy. The normative value of the Nash equilibrium concept should therefore mainly be judged on empirical grounds. It depends crucially on the degree of self-fulfillment of Nash's prophecy in real life.

There has been quite a struggle over refinements of the Nash equilibrium. I only wish to address the *subgame perfect equilibrium* suggested by Reinhard Selten (1965, 1975). To illustrate this refinement, consider the logic of deterrence in an everyday life example. A member of the academic staff receives attractive job offers from other universities and tells the head of department "give me a raise or I'll quit." The head responds by augmenting her salary, because she is needed and he expects her to carry out the threat. In an extensive game model of this interaction it would be easy to see that both individuals are playing best responses to each other's strategies – no reason to refine anything. Now, let us suppose instead that the staff member receives job offers that involve a substantial pay reduction. The Nash equilibrium just described still exists but seems rather odd from a rationality perspective. It is now irrational to carry out the threat, given the pay reduction that will result. Nash's equilibrium properties are only satisfied because the threat will not have to be carried out if every player behaves according to the Nash equilibrium. Selten introduced the concept of subgame perfection to exclude this kind of Nash equilibrium where players have to act against their own interests in certain situations.

Selten's general point is the following. For a solution concept based on careful reasoning it seems inappropriate to assume (a) rationality when players are choosing strategies and (b) irrationality when the same players actually play their strategies. I couldn't agree more but, on the other hand, deterrence is a frequently observed phenomenon and it often works exactly because our evolved psychology makes us stubborn enough to carry out the kinds of threats not allowed by subgame perfection. Emotions play a role in creating this stubbornness and they may even have been shaped by evolution to increase the credibility of threats. In my view, Selten's conceptual refinements (see also Harsanyi and Selten 1988) have demonstrated that an ever more refined "careful reasoning approach" reveals progressively how much of a discrepancy there is between (i) purely theoretical ideas about what it means to make rational decisions, and (ii) scientific insights about the cognitive processes, including reasoning, that take place when real people make their decisions.

## 1.3 GAME THEORY IN REAL LIFE

Classical game theory has strongly inspired theory development in evolutionary biology, but the transfer of ideas suffered initially from the fact
that von Neumann's minimax solution concept received more attention
in textbooks than it deserved at the time (e.g. Luce and Raiffa 1957). The
population geneticist Lewontin (1961), for example, proposed a biological
model in which populations play minimax strategies and thereby ensure
their long-term survival. Lewontin's approach did not spur any further
theory development because it exaggerated the power of natural selection
acting at high levels of aggregation.

Maynard Smith and Price (1973) deliberately avoided stepping into
this pitfall and founded evolutionary game theory in the firm belief that
it would be inappropriate to use the minimax approach (the only one they
knew from classical game theory) for Darwinian studies. It was in the
attempt to create a new concept that they reinvented – as a side effect –
the Nash equilibrium. Maynard Smith and Price gave this equilibrium
an interpretation that very explicitly disentangled game theory from its
rationality assumptions and firmly established its role in biology. The key
idea (see also Maynard Smith 1982) of evolutionary game theory is to
conceive of strategies as animal traits with some genetic heritability and
to search particularly for *evolutionarily stable strategies* (ESS), i.e. strategies that would be maintained by natural selection once they were established in a population. Technically, the ESS concept is a refinement, or
strengthening, of the Nash equilibrium concept, in that it adds a further
condition to Nash's "best response" condition.

In Maynard Smith's framework, games serve to model the interactions
among individuals in a population. These interactions are assumed to
take place with some regularity generation after generation. Game payoffs
are expressed in terms of fitness effects and are used to specify a selection equation. In many models of evolutionary game theory, selection
will carry the population to an ESS that can indeed be found as a Nash
equilibrium of the game (e.g. Hofbauer and Sigmund 1998). Methods
from classical game theory will therefore help biologists find an ESS (e.g.
Hammerstein 2001). Yet the ultimate analysis of evolutionary stability
requires modeling of the evolutionary dynamics and can easily go wrong
if biologists simply appeal to the authority of classical game theory.

A population geneticist may take a skeptical view of evolutionary
game theory because it studies frequency-dependent selection at the
phenotypic level. Focusing one's analytical efforts at this level involves a

substantial risk of overestimating the adaptive power of natural selection. Surely, in real animal populations multiple gene effects, genetic recombination, epistasis, and selfish genetic elements can impede phenotypic optimization. But birds have wings with superb aerodynamic properties. These wings are not cast into quadratic shapes by genetic constraints, as Maynard Smith once put it. In contrast, Karlin (1975) took the equally justified stance that mathematical models of genetic evolution with more than one gene often fail to possess the property of phenotypic optimality at equilibrium. In the same spirit, Moran (1964) had advocated the "nonexistence of adaptive topographies." The mathematical results by Karlin, Moran, and others certainly raised doubt about the "phenocentric" methodology of evolutionary game theory.

But if the theory does not permit the bird to have wings, there must be a problem with the theory. Reinhard Selten and myself introduced the *streetcar paradigm* to resolve the apparent incompatibility between the genetic and the phenotypic level of explanation (Hammerstein and Selten 1994; Hammerstein 1996a). With this paradigm we tried to capture the essence of previous work by Ilan Eshel and Marc Feldman (Eshel and Feldman 1984; see Hammerstein 1996b for details about the history of the ideas). In a nutshell, the paradigm draws attention to the possibility that evolution will change the genetic system when genetic constraints become a strong impediment to phenotypic adaptation.

With genetic rearrangements in mind, evolution can indeed be compared to a journey on a streetcar. The streetcar is a population and its passengers are meant to be the genes in this population. The streetcar moves in phenotype space and in our paradigm we draw attention to the stops (equilibrium states of the evolving population). At each stop new passengers (duplicate genes, genes acquired from other organisms, etc.) enter the streetcar. Sooner or later one or more of the new genetic passengers will resolve the constraints that caused the streetcar to halt at a phenotypically maladaptive stop (equilibrium). This puts the streetcar in motion and the journey continues.

But what if a final stop is reached in which there is no scope for new genes to put the streetcar in motion again? Such a stop will not exist unless we close the ticket counter for certain passengers. Let us therefore exclude as new passengers truly selfish genetic elements and genes that fundamentally alter the character of the phenotypic game. A number of mathematical results, starting with the seminal paper by Eshel and Feldman (1984), suggest that a final stop is necessarily a phenotypically adaptive state (Hammerstein and Selten 1994; Eshel 1996; Hammerstein

1996a; Weissing 1996). Eshel refers to the conceptual background of these results as the "theory of long-term evolution." I am promoting the term "streetcar theory" because it shifts emphasis from *timescales* to fundamental *differences between stops*. Final stops are not final forever but merely better protected against the effects of new genes than temporary stops. In the mathematical theory under discussion little is said about timescales. These scales depend on how easily molecular mechanisms will generate those "magic new genes" that push streetcars into motion again.

Why does the streetcar paradigm reconcile genetic and phenotypic approaches to evolution? Temporary stops depend on genetic detail and this is the domain of population genetics theory. Final stops depend on phenotypic adaptation because genetic rearrangements have removed genetic constraints. It is an empirical question whether the human observer of natural phenomena would see more streetcars halting at final than at temporary stops. I should add that it seems unfeasible to make the argument of genetic constraints against a theoretical research program where the search is for final rather than temporary stops. Conversely, one cannot criticize the emphasis on maladaptive properties if the research program is directed towards temporary stops. Neither genetic nor phenotypic modelers of evolution will be open to objection if they subscribe to the streetcar philosophy; and the discourse between evolutionary biology, game theory, and economics can often be safely conducted with the final stops of the streetcar in mind.

## 1.4 THE POWER OF STRATEGIC ANALYSIS IN BIOLOGY

An interesting difference between classical and evolutionary approaches to game theory is that the former puts the task of strategy choice into the hands of a brain with fictitious superpowers while the latter "outsources" this task and transfers it to the powerful process of natural selection. This is a strong reason to believe that evolutionary game theory has more to say about adaptive strategies in animals and microbes than classical game theory can offer with respect to reasoning-based human strategy choice. While homo economicus fell off his pedestal when challenged by cognitive psychology (e.g. Tversky and Kahneman 1981; Gigerenzer et al. 2011), we actually do find a great deal of economic "wisdom" in the world of organisms. To illustrate this, let us take a look at bacteria, the predominant form of life on earth.

A particular group of bacteria called *Wolbachia* has found a lifestyle that partly resembles that of the well-known cell organelles called mitochondria. Both the bacteria and organelles reside within cells and are typically transmitted via eggs from mothers – but not via sperm from fathers – to sons and daughters. This is how far the similarities go. While mitochondria cooperate with their hosts and act as "power plants" of the cell, the bacteria are rather known as powerful manipulators. Many strains of *Wolbachia* interfere with host reproduction and development in adverse ways, and they usually fail to offer the host any services. Without strategic analysis we would have to view their impact on hosts as a side effect of bacterial presence, quite like the side effect of a drug. But there is a stunning strategic logic behind this bacterial impact. Let us briefly explore this logic through reasoning inspired by game theory, which is of course only a short cut for a more explicit evolutionary analysis.

Remember that, in our example, sperm do not serve as a means of transportation from one host generation to the next. Males are then a dead end for transmission of the bacteria. Anthropomorphically speaking, *Wolbachia* are "buried alive" when sitting in a male. As an intriguing consequence there is no scope for direct selection forcing the bacteria to cooperate with male hosts. Cooperators would not increase their chances of being transmitted. This simple insight was gained astonishingly late in the history of theoretical biology (Cosmides and Tooby 1981).

Let us now continue the analysis. Can we imagine strategically beneficial and biologically plausible ways for the bacteria to reduce the negative prospects of getting stranded in male organisms? Yes, without previous knowledge but inspired by the idea of strategic reasoning, even second-year biology students can figure out that *Wolbachia* could *feminize males* to make them produce eggs instead of sperm, *induce parthenogenesis* to force the exclusive production of daughters, or simply *kill males* if this reduces competition among siblings of the host. In the latter case, not the killer but other individuals of the same bacterial strain are beneficiaries of the manipulation if they live in an infected sister of the killed male. Now, almost too good to be true, all these manipulations have been found in nature (O'Neill et al. 1997; Werren et al. 2008).

Another evolved strategy of *Wolbachia* reveals even more strategic finesse. This strategy employs a twofold trick. Firstly, the bacteria act in male insects like a microbial "pill for males." That is, when sperm from infected males fuse with eggs of uninfected females, this has fatal consequences for early embryonic development. Secondly, when the eggs are infected with the same strain, the bacteria manage to prevent the damage and normal development takes place. At an abstract level, theoretical biologists call this a "poison-antidote system," because the bacteria act as if they poisoned the sperm and provided the antidote in the egg. In an infected population this promotes the infection by reducing the fitness of uninfected females. Even more interesting, the poison-antidote trick will often force infected females to support the infection and foster its transmission to their own offspring (Koehncke et al. 2009). In a figurative sense females then have to pay protection money to *Wolbachia* in order to get their eggs protected from damage through *Wolbachia*. The bacterial strain thus acts like a protection-money racketeer gang.

Obviously, the bacteria do not behave in the sense that the term "behavior" is used for animals and humans. They cannot walk or talk and are not even in possession of a brain. It thus may appear ludicrous to theorize about them with guidance from the behavioral sciences. The heuristic value of strategic analysis, however, is immense. Without the influence of evolutionary game theory we would still consider *Wolbachia* as one of the exhibits in nature's cabinet of curiosities.

I should add that, even without legs, *Wolbachia* are able to move in a controlled way within their hosts. In the egg, for example, the bacteria attach to specific motor proteins of the cell for a ride, like entering a tram. The tram brings them to a specific location that I would describe as the "entry door" to the germ line. Similarly, even without a brain they selectively kill males and leave females alive. Lacking arms, they cannot use an axe to damage sperm, but have found chemical ways of achieving this goal and also of repairing the damage. Human pharmaceutical companies had a hard time achieving anything similar with hundreds or thousands of researchers involved and we biologists are only beginning to understand in mechanistic detail how *Wolbachia* do their trick (Bossan et al. 2011).

## 1.5 GAMES IN THE ANIMAL WORLD: DOES SUBGAME PERFECTION HAVE A CHANCE?

An interesting theoretical and empirical question is whether subgame perfection (already discussed in Section 1.2) plays a role in animal games. Originally this concept requires rational behavior even in those situations that are never reached at a Nash equilibrium. Evolution by natural selection would, of course, have no possibility of generating optimal behavior for specific situations that never occur. This sounds like a clear reason for biologists to dismiss the concept under consideration. Furthermore, Gale et al. (1995) showed for the famous ultimatum game that evolution-like processes could indeed lead to equilibria that are not subgame perfect.

In defense of a modest use of subgame perfection in biological theory (Hammerstein 2001) it must be said, however, that the appropriate way of bringing the concept into biology is through the model of the trembling hand. In this model players make mistakes in executing their strategies. Moves that are not part of a strategy are chosen by mistake with a small probability. Selection can now, in principle, adapt behavior to those situations that originally would not have been reached. In practice, i.e. in finite populations, this requires error rates large enough to give selection a bite. The original trembling hand approach considers vanishingly small error rates and this is the aspect we do not take literally in biological theory. Having said this, there are many problem areas where it is intellectually fruitful to focus attention on perfect equilibria.

To illustrate this, consider the parental investment game between a male and a female. They have copulated and their offspring will need some care. Who is going to provide the care? Suppose we are studying an animal example in which both male and female would benefit from letting only their partner care for the offspring. Suppose further that the male has established a mating territory in which he has to stay to attract as many partners as possible, whereas the female has no a priori reason to remain in this confined area. The situation then resembles an ultimatum game where she is the proposer and her male mate the responder. The reason for these asymmetric roles is that she can leave him alone with the fertilized eggs, whereas he will have the offspring in his territory and no incentive to leave. If she deserts, the male can either "accept" the deal and pay the price for his parental effort, or "reject" and let the offspring die. But would evolution permit this threat of a passive infanticide?

To answer this question, let a population play the imperfect Nash equilibrium where the male would ignore his offspring and the female would stay and care. Let us make the realistic assumption that occasionally a predator, lightning, or other environmental effect kills the female. These events can play the role of the tremble in the trembling hand approach because the female is programmed to care but instead leaves the father alone with the offspring. If this happens frequently enough, compensatory male care will evolve under a wide range of assumptions. In a subsequent evolutionary step, deliberate mate desertion will evolve, females will always disappear and males will routinely "look after the babies." At this point, an evolutionary transition towards the subgame perfect equilibrium has taken place, imposing the burden of parental care on the male. The idea of perfection has helped us understand the nature of an interesting biological problem.

One could list a plethora of biological examples that can be idealized through the abstract scenario just described, and where males actually do carry the burden of parental care. Many of these examples are found in bony fishes with external fertilization and, for example, in insect species related to the giant water bug (*Adebus herberti*). The giant water bug itself teaches us how evolution can further shape the details of parental care. Here, females lay their fertilized eggs literally on the male's back (wings). The male then carries them around until they hatch (Smith 1979). At this advanced stage in the evolution of paternal care, male water bugs cannot even mate with other females while caring for their offspring.

What about the ultimatum game played in the labs of experimental economists? Güth et al. (1982) demonstrated that – unlike water bugs – humans practically ignore the subgame perfect equilibrium in this game. Henrich et al. (2005, 2010) conducted anthropological studies and took the ultimatum game to fifteen cultures. They found essentially the same phenomenon but also variation within and between cultures in how strongly people deviate from the perfect equilibrium. Fairness issues play a role but it is difficult to disentangle biological and cultural factors in this context (Hagen and Hammerstein 2006).

Finally, let us come back to the general problem of deterrence already discussed in Section 1.2. It seems clear that deterrence can frequently be observed in the human world, that it often works, and that in many cases it seems to violate the idea of subgame perfection. Hirshleifer (1977) argued that in order to understand deterrence we have to pay strong attention to the emotions involved. From his perspective, anger and other emotions can act as *internal* commitment devices. Anger in some sense automatizes the execution of a threat, just as if one had installed a spring gun. Darwinian decision theory will now have to ask, whether any emotion evolved *because* it was an internal commitment device. The answer probably is not as simple as that.

## 1.6 A CONCLUDING REMARK ON LEARNING AND THE HUMAN ANIMAL

Most biologists consider humans on the one hand as "yet another animal species" but admit on the other hand that we are rather special. What is so special about us from an evolutionary point of view? We are particularly well equipped with prerequisites for social life and possess an outstanding capability for cooperation. This capability seems unmatched by any other animal species, especially once one takes into account that human cooperation frequently occurs among people who are not close genetic relatives.

While many scholars in the humanities strongly emphasize the cultural influence on our behavior and decision making, evolutionary anthropologists are interested in the ways evolution has shaped the patterns by which this influence occurs. The evolution of social learning then becomes the major issue (Boyd and Richerson 1985; Boyd et al. 2011). Learning is always prepared by evolution in the sense that animal species differ strongly in what they can learn and how they do it. The human learning machinery is no exception. It is evolutionarily designed to be efficient in a social environment (see Hammerstein and Boyd, in press, for a review). Evolutionary biology, therefore, matters in the study of learning despite the fact that the indisputable predominance of learning is often thought to sound the death knell for a biological approach to the study of human behavior.

Learning also played a role when theoreticians in economics transferred evolutionary game theory from biology to their own discipline. For the reasons given above, they had to rely on social learning rather than genetic evolution as the process that generates strategic behavior. Without knowing the biological theory under discussion, John Nash had already foreseen this possibility. But his visionary statements were buried in his unpublished PhD thesis, not mentioned in any textbook, and only rediscovered when he received the Nobel Prize (Kuhn et al. 1995).

The amazing similarity between evolutionary game theory in biology and economics has led many of us to believe that learning and evolution are two very similar processes that just happen to operate on different timescales. As a first approximation this is indeed a good idea and has fostered the interdisciplinary dialogue. In the long run, however, we should not close our eyes to a number of important differences. The transmission of genetic material, on the one hand, is a highly mechanistic process that can easily be cast into equations. We know, for example, that genes are only transmitted from parents to offspring. Social learning, on the other hand, can occur in a variety of different ways. The learners can differ in whom they imitate, what they imitate, and how much they rely on their own experience. Social learning can take place under different emotional states and in interference with higher cognitive processes. Confronted with this complexity it seems difficult to achieve conceptual clarity. But since evolution has had a hand in it, there are probably hidden regularities that will permit us to develop a Darwinian theory of learning and decision making – an agenda for the future.

## REFERENCES

Bossan, B., Koehncke, A., and Hammerstein, P. (2011). A new model and method for understanding *Wolbachia*-induced cytoplasmic incompatibility, *PLoS ONE*, 6(5), e19757.

Boyd, R. and Richerson, P. J. (1985). *Culture and the Evolutionary Process.* University of Chicago Press.

Boyd, R., Richerson, P. and Henrich, J. (2011). The cultural niche: Why learning is essential for human adaptation. *PNAS*, 108, 10918–10925.

Cosmides, L. M. and Tooby, J. (1981). Cytoplasmic inheritance and intragenomic conflict. *Journal of Theoretical Biology*, 89, 83–129.

Eshel, I. (1996). On the changing concept of population stability as a reflection of a changing point of view in the quantitative theory of evolution. *Journal of Mathematical Biology*, 34, 485–510.

Eshel, I. and Feldman, M. W. (1984). Initial increase of new mutants and some continuity properties of ESS in two locus systems. *American Naturalist*, 124, 631–640.

Gale, J., Binmore, K. G., and Samuelson, L. (1995). Learning to be imperfect: The ultimatum game. *Games and Economic Behavior*, 8, 56–90.

Gigerenzer, G., Hertwig, R., and Pachur, T., eds. (2011). *Heuristics: The Foundations of Adaptive Behavior.* New York: Oxford University Press.

Güth, W., Schmittberger, R., and Schwarze, B. (1982). An experimental analysis of the ultimatum game. *Journal of Economic Behavior and Organization* 3, 367–388.

Hagen, E. H. and Hammerstein, P. (2006). Game theory and human evolution: A critique of some recent interpretations of experimental games. *Theoretical Population Biology*, 69, 339–348.

Hammerstein, P. (1996a). Darwinian adaptation, population genetics and the streetcar theory of evolution. *Journal of Mathematical Biology*, 34, 511–532.

(1996b). Streetcar theory and long-term evolution. *Science*, 273, 1032.

(2001). Economic behaviour in humans and other animals. In *Economics in Nature*, eds. R. Noë, J. A. R. A. M. van Hooff, and P. Hammerstein, 1–19. Cambridge University Press.

Hammerstein, P. and Boyd, R. (in press). Learning, cognitive limitations and the modeling of social behavior. In *Evolution and the Mechanisms of Decision Making*, eds. P. Hammerstein and J. R. Stevens. MIT Press.

Hammerstein, P. and Selten, R. (1994). Game theory and evolutionary biology. In *Handbook of Game Theory with Economic Applications*, vol. 2, eds. R. J. Aumann and S. Hart, 929–993. Amsterdam: Elsevier.

Harsanyi, J. and Selten, R. (1988). *A General Theory of Equilibrium Selection in Games.* Cambridge, MA: MIT Press.

Henrich, J., Boyd, R., Bowles, S., Camerer, C., Fehr, E., Gintis, H., McElreath, R., Alvard, M., Barr, A., Ensminger, J., Hill, K., Gil-White, F., Gurven, M., Marlowe, F., Patton, J. Q., Smith, N., and Tracer, D. (2005). "Economic man" in cross-cultural perspective: Behavioral experiments in 15 small-scale societies. *Behavioral and Brain Sciences*, 28, 795–855.

Henrich, J., Ensimger, J., McElreath, R., Barr, A., Barrett, C., Bolyanatz, A., Cardenas, J. C., Gurven, M., Gwako, E., Henrich, N., Lesorogol, C., Marlowe, F., Tracer, D., and Ziker, J. (2010). Markets, religion, community size, and the evolution of fairness and punishment. *Science*, 327, 1480–1484.

Hirshleifer, J. (1977). Economics from a biological viewpoint. *Journal of Law and Economics*, 20, 1–52.

Hofbauer, J. and Sigmund, K. (1998). *Evolutionary Games and Population Dynamics*. Cambridge University Press.

Karlin, S. (1975). General two-locus selection models: Some objectives, results and interpretations. *Theoretical Population Biology*, 7, 364–398.

Koehncke, A., Telschow, A., Werren, J. H., and Hammerstein, P. (2009). Life and death of an influential passenger: *Wolbachia* and the evolution of CI-modifiers by their hosts. *PLoS ONE*, 4(2), e4425.

Kuhn, H. W., Harsanyi, J. C., Selten, R., Weibull, J. W., van Damme, E., Nash, J. F., and Hammerstein, P. (1995). The work of John F. Nash Jr. in game theory: Nobel Seminar, December 8, 1994. *Duke Journal of Mathematics*, 81, 1–29.

Lewontin, R. C. (1961). Evolution and the theory of games. *Journal of Theoretical Biology*, 1, 382–403.

Luce, R. D. and Raiffa, H. (1957). *Games and Decisions*. New York: Wiley.

Maynard Smith, J. (1982). *Evolution and the Theory of Games*. Cambridge University Press.

Maynard Smith, J. and Price, G. R. (1973). The logic of animal conflict. *Nature*, 246, 15–18.

Moran, P. A. P. (1964) On the nonexistence of adaptive topographies. *Annals of Human Genetics*, 27, 383–393.

Nash, J. F. (1951). Non-cooperative games. *Annals of Mathematics*, 54, 286–295.

O'Neill, S. L., Hoffmann, A. A., and Werren, J., eds. (1997). *Influential Passengers: Inherited Microorganisms and Arthropod Reproduction*. Oxford University Press.

Savage, L. J. (1954). *The Foundations of Statistics*. New York: Wiley.

Selten, R. (1965). Spieltheoretische Behandlung eines Oligopolmodells mit Nachfrageträgheit – Teil I: Bestimmung des dynamischen Preisgleichgewichts. *Zeitschrift für die gesamte Staatswissenschaft*, 121, 301–324.

   (1975). Reexamination of the perfectness concept for equilibrium points in extensive games. *International Journal of Game Theory*, 4, 25–55.

   (1991). Evolution, learning, and economic behavior. *Games and Economic Behavior*, 3, 3–24.

Smith, R. L. (1979). Paternity assurance and altered roles in the mating behaviour of a giant water bug, *Abedus herberti* (Heteroptera: Belostomatidae). *Animal Behaviour*, 27, 716–725.

Tversky, A., and Kahneman, D. (1981). The framing of decisions and the psychology of choice. *Science*, 211, 453–458.

von Neumann, J. and Morgenstern, O. (1944). *Theory of Games and Economic Behavior*. Princeton University Press.

Weissing, F. J. (1996) Genetic versus phenotypic models of selection: Can genetics be neglected in a long-term perspective? *Journal of Mathematical Biology*, 34, 533–555.

Werren, J. H., Baldo, L., and Clark, M. E. (2008). *Wolbachia*: Master manipulators of invertebrate biology. *Nature Reviews Microbiology*, 6, 741–751.

# CHAPTER 2

# What do humans maximize?

*Claire El Mouden, Maxwell Burton-Chellew, Andy Gardner
and Stuart A. West*

## 2.1 INTRODUCTION

Natural selection results in organisms that appear designed to maximize their inclusive fitness (Figure 2.1; Hamilton 1964, 1970). Our null hypothesis is therefore that people, like all organisms, behave in ways which reveal that their ultimate goal is inclusive fitness maximization. Saying that people appear designed to maximize their inclusive fitness does not imply that they are aware of this design objective or that they will achieve maximum possible inclusive fitness. Evolutionary theory does not predict that humans will intentionally try to maximize anything. This statement does not imply an absence of individual autonomy of action; natural selection has not hard-wired all of our behaviours and decisions. However, in large part, it has defined which things cause people to experience pleasant or unpleasant emotions and sensations. For example food, sex, friendship, being of use to others, a sense of security and social recognition are enjoyable, while hunger, pain, fear, failure and ostracism are unpleasant. These preferences mean that people derive pleasure from, and direct their behaviour toward, evolutionarily beneficial outcomes. In short, humans are largely free to do what they want (they can choose how to satisfy their desires), but they are not free to want what they want (their desires are shaped by evolved preferences).

Humans have an unrivalled capacity for problem solving, reasoning and both individual and social learning. These cognitive tools allow individuals to use their experiences to follow a unique path toward happiness throughout their lives. If people have different experiences to learn from, they will behave differently so we expect people of different age groups, sexes, institutions, classes, communities and cultures to behave differently and to hold divergent beliefs about how to achieve the same goals.

While human preferences may be the result of natural selection, there are good reasons to expect people to make mistakes. There is

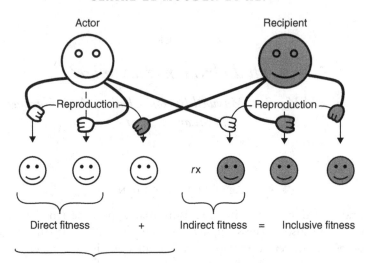

Figure 2.1.  Inclusive fitness is the sum of an individual's direct and indirect fitness (Hamilton 1964). Social behaviours affect the reproductive success of self and others. The impact of the actor's behaviour (white hands) on its own reproduction (white offspring) is the direct fitness effect. The impact of the actor's behaviour (white hands) on the reproductive success of social partners (grey offspring), weighted by the relatedness of the actor to the recipient, *r*, is the indirect fitness effect. Inclusive fitness does not include all of the reproductive success of relatives (grey offspring), only that which is due to the behaviour of the actor (white hands). Also, inclusive fitness does not include all of the reproductive success of the actor (white offspring), only that which is due to its own behaviour (white hands). A key feature of inclusive fitness is that, as defined, it describes the components of reproductive success which the actor can influence, and therefore what they could be appearing to maximize.
*Source:* adapted from West et al. 2007b.

no expectation that natural design will be perfect or optimal as adaptation is constrained by factors such as variable selection pressures, fitness trade-offs and environmental change. As a result, humans will not always behave in ways that maximize their inclusive fitness, or anything else such as happiness or income. As natural selection does not maximize happiness (from a Darwinian perspective, happiness is not the goal, but a means to an inclusive-fitness-maximizing end), or proxies for happiness such as income, there is no expectation that humans would always match the predictions of models that assume they do. Rather, natural selection creates our desire for happiness in order to manipulate our choices toward adaptive behaviour. Therefore, while economic models are extremely useful for studying human behaviour, we suggest there is little point testing whether humans behave optimally as fitness/happiness/income maximizers (we know they will not). Instead, we advocate the biological approach, which uses economic tools to generate predictions that provide a framework to help us study human decision making and to understand why humans show the adaptations they do.

In the next part of this chapter, we discuss the purpose of adaptation and the reasons why organisms are expected to be neither perfect nor optimal. Then, in the second part, we go on to ask 'what does evolutionary theory predict humans will maximize?' In the third section, we discuss the reasons why human actions sometimes appear not to maximize inclusive fitness (or anything else) and finally, given these reasons, we discuss when humans will match the predictions of optimality models.

## 2.2 WHAT IS THE PURPOSE OF ADAPTATION?

In crossing a heath, suppose I pitched my foot against a *stone*, and were asked how the stone came to be there, I might possibly answer that, for anything I knew to the contrary, it had laid there for ever ... But suppose I had found a *watch* upon the ground, and it should be enquired how the watch happened to be in that place, I should hardly think of the answer which I had before given, that, for any thing I knew, the watch might have always been there. Yet why should not this answer serve for the watch, as well as for the stone?

The opening lines of *Natural Theology* (Paley 1802, p. 7).

What is the difference between a watch and stone? Unlike stones, Paley explains that watches consist of many parts, and that the form of these individual parts may only be understood when we realize they are all contrived for a common purpose (in this case to tell the time). In other words, unlike stones, watches appear designed. Paley argues that organisms resemble watches, not stones by demonstrating that the parts of an organism are like watches, made up of related parts, contrived for common purposes. Consider the eye; the function of an eye's retina, cornea, lens, muscles and nerves cannot be understood unless we appreciate these parts are contrived for the common purpose of seeing. So when Paley asked 'what is the difference between a watch and a stone?', he had identified the question which lies at the heart of evolutionary biology: why does the natural world appear designed (Darwin 1859; Williams 1966; Leigh 1971; Maynard Smith 1982; Gardner 2009)? Fifty years later, Darwin provided the answer.

Darwin's (1859) theory of natural selection accounts for both the *process* and the *apparent purpose* of adaptation (Gardner 2009). The process is simple: as more individuals are born than survive to reproductive age, any heritable traits associated with increased individual reproductive success will tend to accumulate in populations. As a consequence of this process, Darwin argued, successive generations of organisms will appear increasingly well designed to maximize their reproductive success. Darwin therefore answered Paley's question in two ways: first, he showed how complex design (adaptation) could result from a natural process; second, he explained what natural design was ultimately for – to maximize Darwinian fitness.

The *Origin of Species* is a masterpiece, but its central argument is informal (non-mathematical) and it was written before the mechanism of biological inheritance (genes) was known. This meant Darwin's ideas were not precisely defined. In the 1930s, Fisher formally linked the dynamics of population genetics to the process of natural selection in his seminal work *The Genetical Theory of Natural Selection* (Fisher 1930). In this work he proposed his 'Fundamental Theorem of Natural Selection', which formally defines natural selection in terms of changes in gene frequencies and defines Darwinian fitness as an individual's genetic contribution to future generations. His theorem shows that genes associated with a greater individual fitness are predicted to accumulate in natural populations. Given this result, Fisher concluded that organisms would appear as if they are striving to maximize their Darwinian fitness (see also Grafen 2002, 2007).

Fisher's theory contained an important caveat; while he knew that behaviours may be favoured because of their indirect effect on relatives who share genes, he explicitly chose to ignore this added complexity. It was Hamilton (1964, 1970) who first formally incorporated the effect of relatives into fitness, and thus made what is arguably the most important contribution to evolutionary theory since Darwin. The result was inclusive fitness theory, which provides a more complete understanding of the process of natural selection, and is the modern view of what natural selection maximizes. An individual's inclusive fitness may be divided into two components (Figure 2.1). The first component is direct fitness, which is a measure of an individual's genetic contribution to future generations via its own reproduction. The second component is indirect fitness, which measures the genetic contribution an individual achieves by affecting the reproduction of related individuals. Grafen (2002, 2006, 2007) has confirmed Hamilton's result more formally, by demonstrating that the phenotypes resulting from the dynamics of natural selection on gene frequencies in a related population match the result of an optimization program where an agent is set to maximize its inclusive fitness within a phenotypic strategy set. Put simply, the fitness that organisms should appear designed to maximize is inclusive fitness.

Table 2.1. *A Hamiltonian classification of social behaviours.*

| Effect on actor (direct fitness) | Effect on recipient (indirect fitness) | |
|---|---|---|
| | + | – |
| + | Mutually beneficial | Selfish |
| – | Altruistic | Spiteful |

A social behaviour refers to any trait that has fitness consequences for another. A Hamiltonian (or evolutionary) classification is determined by the effect the behaviour has on the individual's average lifetime fitness and not by the fitness consequences of single interactions. Both mutually beneficial (+/+) and selfish (+/–) traits may evolve between relatives or non-relatives. Altruistic behaviours (–/+) are favoured when the benefits to related individuals outweigh the cost to self. Spiteful traits (–/–) are very rare, because of the need for negative relatedness (where the actor and recipient on average share fewer genes than two individuals drawn at random from the population).

This expanded view of fitness was instrumental in explaining the diversity of sociality in the natural world (Hamilton 1996). A behaviour is cooperative if it has been selected for, at least in part, because of the beneficial effect it has on others (West et al. 2007b). Explaining the presence of such behaviours appears problematic given that – all else being equal – they reduce the relative fitness of the actor. However, inclusive fitness theory showed that natural selection could favour cooperation or the limitation of conflict under a wide range of conditions. One explanation is that seemingly disadvantageous genes can increase their transmission indirectly by helping other individuals (typically close relatives) that are likely to share the same gene (Hamilton 1964). Yet cooperation also occurs between unrelated individuals and even between different species.

The inherent instability of cooperation between non-relatives is often conceptualized with the aid of the prisoner's dilemma (Axelrod and Hamilton 1981) or the tragedy of the commons (Hardin 1968), whereby individuals do best by not cooperating (cheating), no matter what their partners do. When there is no scope for repeated interactions or sanctions, this results in an inevitable outcome (the 'tragedy') in which all rational actors cheat, even though they would have been better off if they had all cooperated, hence the dilemma. Thus for cooperative behaviour between non-relatives to be evolutionarily stable, it must be favoured by hidden direct fitness benefits that outweigh any apparent costs and, when

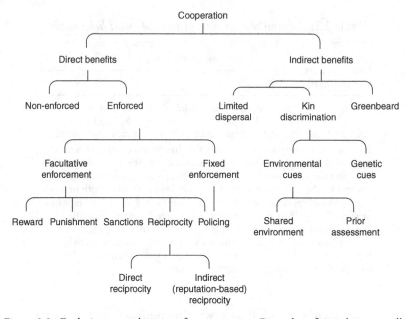

Figure 2.2. Evolutionary explanations for cooperation. Direct benefits explain mutually beneficial cooperation, which may be between non-relatives, whereas indirect benefits explain altruistic cooperation, which is always between relatives (Hamilton 1964). Within these two fundamental categories, different mechanisms can be classified in various ways (Frank 2003; Sachs et al. 2004; Lehmann and Keller 2006; Bergmüller et al. 2007; West et al. 2007b). This classification is intended for illustration only. Not all explanations are included (e.g. group competition, group augmentation and the effect of demography are excluded). A single act of cooperation may require multiple explanations; for example, it may evolve due to both direct and indirect fitness benefits, or interactions with relatives could be maintained by both limited dispersal and kin discrimination. Furthermore, it does not show how the relative importance of different mechanisms may shift through time. For example, group members may need to be related for a group to form so the benefits of cooperation are sufficient; but once established, cooperation may remain stable even if group members are unrelated, due to enforcement mechanisms. In most social species, different explanations will apply in different circumstances. For example, in humans, shared environment mechanisms help explain cooperation between kin, whereas reputation and punishment are important for cooperation between non-relatives. Finally, the way we divide up facultative enforcement strategies here is overly simplistic; a detailed discussion is beyond the scope of this chapter, and is provided elsewhere (Bergmüller et al. 2007). *Source:* adapted from West et al. 2007b.

confronted with a new instance of sociality, it is the job of evolutionary biologists to elucidate these (Sachs et al. 2004; Lehmann and Keller 2006; West et al. 2007b).

Over the past forty years, an extensive literature has developed within evolutionary biology, building upon Hamilton's work, to explain how natural selection may favour the formation and maintenance of cooperation at all levels of biological complexity, including between non-relatives and for different demographic and population structures (Table 2.1; Figure 2.2). Broadly speaking, such cooperation may be favoured for one or other of two reasons: because it is also directly beneficial in the long run to the performer, more so than it is costly (as, for example, in group augmentation effects; Kokko et al. 2001); or because it is enforced, through policing (Frank 1995, 2003; El Mouden et al. 2010), punishment (Clutton-Brock and Parker 1995; Gardner and West 2004; Lehmann et al. 2007) or sanctioning of cheaters (Kiers et al. 2003), and/or through rewards to cooperators, via mechanisms such as direct or indirect reciprocity (Trivers 1971). This work illustrates that individuals are not born selfish (contra Dawkins 1976, p. 3): they are born as inclusive fitness maximizers and this may favour their being selfish, spiteful, mutualistic or altruistic, depending on the circumstances in which they find themselves (West et al. 2007a).

## 2.3 ADAPTATION DOES NOT IMPLY PERFECTION OR OPTIMALITY

Watches vary. A stick in the ground can make a crude sundial, accurate to the nearest hour or so on a sunny day, while an atomic clock, cooled to near absolute zero, has a margin of error of about 1 second in 30 million years. Nevertheless, both sundials and atomic clocks are designed to tell the time. Even a broken watch that does not work at all shows evidence of design. According to Paley: 'It is not necessary that a machine be perfect, in order to shew with what design it was made ... the only question is, whether it were made with any design at all' (Paley 1802, p. 8). Applied to nature, Paley's point is that living organisms need not be perfect or optimal fitness maximizers to exhibit adaptations for 'fitness-maximizing' design (Gardner 2009).

The closest thing to perfect timekeeping is the atomic clock. If natural selection could produce its version of a 'perfect' watch, it would be a 'Darwinian Demon' (Law 1979). This imagined organism could simultaneously maximize all aspects of its inclusive fitness – it would live forever, in any environment, eat anything, be able to reproduce as soon as it was born and give birth to high-quality offspring at an infinite rate while also helping its relatives produce their offspring at an infinite rate. Darwinian demons are a fantasy, for the same reason that we cannot wear atomic clocks on our wrists – constraints. In nature, phylogenetic, physical, time, developmental, information and resource constraints mean that organisms are unable to be perfect fitness maximizers. In addition to constraints, organisms must efficiently allocate limited resources to competing concerns. Therefore, from an evolutionary perspective, organisms can never be perfect. However, it is worth asking: when organisms make their complex trade-offs between different aspects of fitness (e.g. growth vs own reproduction vs relatives' reproduction), can they do so optimally?

Optimality is the achievement of maximum fitness given the constraints an organism faces. This means both atomic clocks and sundials may represent optimal designs in different environments. However, just as a broken watch still shows evidence of design, the presence of fitness-maximizing adaptation does not imply optimality of design. This is because in natural populations, in addition to genes, many other factors influence an organism's chances of survival and reproduction (i.e. its fitness). As the opening lines of Fisher (1930) famously read 'Evolution is not natural selection.' Stochastic effects (e.g. unpredictable weather), mutation and population movements all shift gene frequencies, so these are evolutionary forces too. If selection pressures are weak (most individuals survive and reproduce), these non-selective evolutionary forces can be the main drivers of phenotypic change, not natural selection. As these forces do not maximize anything, they divert phenotypes away from natural selection's goal of optimality, degrading any appearance of design. Therefore, even though organisms can be optimal, we expect measurably optimal behaviour to be extremely rare. There is no expectation that organisms will reach the goal of optimality, or even be anywhere near it, just that natural selection will direct genetic variation in that direction.

Although adaptation does not imply optimality, it is important to note that mechanisms under strong selection, such as those concerning life or death decisions or commonly encountered situations, will be more finely tuned toward optimality, as here, natural selection will be the dominant evolutionary force. Thus we expect to see organisms performing 'better' in the situations that they encounter most frequently. For example, as sex ratio theory predicts, a fig wasp will vary the sex ratio of her offspring depending upon the number of other females that lay eggs in the same fruit (Herre 1985, 1987). Studies have demonstrated that the sex ratios the wasps produce are closer to the predictions of evolutionary optimality models in the situations they, as a population, encounter more frequently (Herre 1987).

## 2.4 WHAT DOES EVOLUTIONARY THEORY PREDICT HUMANS WILL MAXIMIZE?

Many animals have evolved the capacity for phenotypic plasticity in order to cope with variable environments. There are various ways phenotypic plasticity can be achieved. Natural selection may favour a range of phenotypes (Smith and Skúlason 1996; Halama and Reznick 2001). For example, many fish species have evolved discrete morphs, which feed in different habitats or adopt different mating strategies (Taborsky 1994; Ehlinger 1989). Alternatively, natural selection may favour the use of proximate cues to help determine optimal behaviour. For example, female great tits use increasing day length, temperature, food availability and social stimulation to time their egg laying with the peak emergence of caterpillars in spring. Studies of a great tit population in Oxford show that since the 1960s, great tits have used these cues to successfully track changes in the temporal availability of caterpillars (Dawson 2008). However, the main way animals achieve behavioural flexibility is through learning.

Learning occurs when an organism changes its behaviour in response to experience (Pearce 1997). By understanding and remembering the consequences of their actions, animals can react better to the behaviour of others and can cope with changes in their local environment (Sullivan 1988; Hollis et al. 1997; Mahometa and Domjan 2005). Many animals exhibit surprising abilities at learning. For example, goldfish, chickens, horses, cats and rhesus monkeys learn equally well to discriminate between two stimuli to gain a reward (Warren 1965). Despite the fact that many behaviours are not hard-wired, an animal's actions will still appear designed to maximize inclusive fitness as its learning is driven by cognitive mechanisms such as hunger, libido, pain and fear which are adapted to direct behaviour toward fitness-maximizing goals.

Humans differ from other organisms by degree not by kind. People derive pleasure from, and direct their efforts toward, fitness-maximizing outcomes (Darwin 1871). For example, people enjoy food, sex, friendship and social recognition and they dislike hunger, pain, fear, failure and ostracism. The diverse and growing literature on human behaviour illustrates that adaptive preferences underpin many aspects of our lives. We summarize their effect to be that overall, humans strive to increase personal 'pleasure' and minimize 'pain', although what gives a person pleasure and pain changes often. We would predict that people are most happy when their conditions are improving, and are less happy when their conditions stagnate, even if they are doing well. This is because the positive emotions associated with happiness are evolved proximate mechanisms to guide our behaviour toward adaptive ends.

To illustrate how adaptive preferences can influence a person's behaviour, we use the literature on sexual preferences as an example. This work shows that men prefer plump women to thin women (81% of people in fifty-eight cultures surveyed; Brown and Konner 1987), and that this preference becomes more extreme when men are hungry (Swami and Tovee 2006), indicating the adaptive explanation may relate to the risk of food shortage (Sugiyama 2004). Men also prefer women with a low waist-to-hip ratio (Singh 1993), which may be adaptive as a low waist-to-hip ratio correlates with offspring with higher cognitive ability (Lasseka and Gaulin 2008). During the few days of the month when women are fertile, they are considered more attractive (Roberts et al. 2004) and male partners are more attentive (Haselton and Gangested 2006) and possessive (Gangestad et al. 2002). While fertile, women prefer more masculine males (Penton-Voak, et al. 1999; Gangestad et al. 2005), make more effort to dress attractively (Haselton et al. 2007) and are more flirtatious and sociable (Haselton and Gangestad 2006). Finally, ovulating lap dancers earn, on average, $30 an hour more than menstruating lap dancers (Miller et al. 2007).

What sets humans apart from all other species is the presence of accumulated cultural knowledge. Culture is defined as information capable of affecting individuals' phenotypes which they acquire from other conspecifics by teaching or imitation (Richerson and Boyd 2004). Many non-human species, including blue tits (Perrins 1979), chimpanzees (McGrew 2004) and dolphins (Simões-Lopes et al. 1998), have simple systems of cultural transmission where non-genetic, non-species-wide behaviours are passed across the generations. However, only humans have the cognitive capacity to support a cumulative culture where skills are learnt, improved and transmitted to others via teaching or imitation (Boyd and Richerson 1985, 2005a; Henrich 2005, 2008). The positive fitness effect of cumulative culture is evident in even simple human technologies such as a 60,000-year-old stone-tipped spear (Pettit 2005). Its design is the product of multiple innovations to the shaft, hafting and point. No modern human could ever arrive in the Savannah and design such a spear on the spot. Unlike other animals, humans rely upon the accumulated cultural knowledge of past generations for survival.

Humans show many adaptations for acquiring and using cultural knowledge. We are unusually docile (Simon 1990), highly sensitive to expressions of approval and disapproval by parents and peers (Baum 1994; Henrich and McElreath 2003) and most importantly, we excel at high-fidelity imitation (Tomasello 1999). These imitation skills enable humans to perform social learning via imitation and practice (Whiten and Ham 1992; Heyes and Galef 1996), which ethnographic studies confirm is how we acquire the majority of our skills (as opposed to doing so independently via individual learning, which is what animals typically do; Fiske 1999). Furthermore, our cognition adapts to the local environment as some behaviours learnt during our childhood can become hardwired in the brain, permanently modifying our minds at the subconscious level (Quartz and Sejnowski 2000; Quartz 2002). This is evidenced by between-cultural variation in susceptibility to optical illusions, hand–eye coordination and male stress and aggression levels (Segall et al. 1966; Cohen et al. 1996; Henrich 2008). It appears that such traits become permanently fixed by about twenty years of age, and remain fixed for life, even if the individual migrates into a new culture (Segall et al. 1966).

Compared with other animals, humans possess cognitive adaptations for problem solving, strategic thought, and behavioural flexibility which are unique. People learn from their life experiences and the experiences of those around them and then use this knowledge to decide how best to seek out happiness and avoid pain. As people have different life experiences, we expect them to behave very differently and to possess widely divergent beliefs about how to fulfill the same evolved preferences. This means that behaviours that appear drastically different may share a common evolutionary foundation. For example, some young men may disrespect their parents, smoke and drive fast cars; other young men may choose to socialize with elder family members, work long hours and avoid contact with girls. Both may be trying to fulfill their desire for social status, but they have learnt to achieve the same goal in very different ways. Similar explanations can sometimes be applied to those that vow to be celibate or to forego material wealth – behaviours often cited as a challenge to evolutionary explanations for human behaviour. (This also ignores the added complication that such 'maladaptive' behaviours may benefit familial interests and so could help to increase relatives' fitness.) Since members of the same family, class, culture or tribal group share many formative experiences in common, it follows that the variation in beliefs within such groups should be far less than the variation found between them.

Humans, like other group-living vertebrates, show adaptations for sociality. This is evidenced by many elegant behavioural economic experiments, which show that we are often willing to cooperate and to punish those who do not cooperate. Importantly, this includes cooperation or punishment toward strangers in one-shot experimental encounters where there is no way of benefiting from the act (Fehr and Gächter 2002; Fehr and Fischbacher 2003; Gintis et al. 2005). Not surprisingly, such behaviour has a neurological basis: the brain's reward centres (e.g. the striatum; Stanfey 2007) activate and make us feel good when we donate money to charity (Moll et al. 2006), observe charitable behaviour (Harbaugh et al. 2007), engage in reciprocal cooperation (Rilling et al. 2002) and in costly punishment (de Quervain et al. 2004). There are many possible evolutionary explanations for why we are motivated to perform these social behaviours (Figure 2.2) and a detailed discussion is beyond the scope of this chapter (see Sachs et al. 2004; Lehmann and Keller 2006; West et al. 2007b, 2011). The fact that these behaviours are seen as irrational (with respect to an income-maximizing agent) in economic experiments involving positive externalities has generated the false impression that humans are uniquely social organisms. Such an impression feeds from misconceptions about evolutionary theory, the difference between ultimate and proximate questions, and the ways that cooperation evolves (West et al. 2011). Indeed, if the recent history of experimental social sciences had instead tested for income-maximizing behaviour in experiments with negative externalities or direct benefits to cooperation, then the same logic would have resulted in a view of humans as irrationally anti-social (Kümmerli et al. 2010).

To explain the adaptive value of social behaviours, we must understand the evolutionary trade-offs and constraints the individual experienced in the environment where the behaviours were selected for and were maintained (Herre 1987). As we do not know the full details of our ancestral environment, it is not possible to know whether social adaptations we now see were favoured due to direct or indirect benefits. Therefore, we will probably never be able to quantitatively measure the extent to which these adaptations for sociality are (from an evolutionary perspective) mutually beneficial or altruistic. However we can glean insights into the key selective pressures that operated in the past with experiments that utilize implicit cues of key environmental features (Tooby and Cosmides 1990). For example, in experiments eye-spots have proved to be a salient cue of reputational effects, and people condition their cooperation levels on the presence or absence of such cues (Bateson et al. 2006; Haley and Fessler 2005; Rigdon et al. 2009; Ernest-Jones et al. 2011).

## 2.5 WHY DO OUR ACTIONS SOMETIME APPEAR NOT TO MAXIMIZE INCLUSIVE FITNESS (OR ANYTHING ELSE)?

Behavioural ecologists commonly use the economic tools of optimization theory and game theory, which assume organisms behave as optimal fitness-maximizing agents (Davies et al. 2012). Constructing such models entails making assumptions about the constraints and trade-offs facing the individual in order to define the available choices (the strategy set) and the potential payoffs (in terms of fitness) they may receive. Crucially, in order to be testable, the model requires feasibly measurable parameters, and because fitness is near impossible to measure, this means that proxies for fitness (e.g. number of offspring, food intake, number of matings or income) and fitness-based payoffs (e.g. cost = time spent on task, calories used or price; benefit = calories gained, matings, change in social rank or profit) must be used (Parker and Maynard Smith 1990). If the data and model match, the model may correctly capture the adaptive purpose of the behaviour being studied. However, often observations do not fit model predictions and it appears that the individuals are not maximizing their inclusive fitness (or anything else). Indeed, while data and models can match well, meta-analyses show that on average, evolutionary or ecological models explain only 2–5% of the variation between natural populations (Møller and Jennions 2002). There are many reasons for this – here we discuss six of them.

First, natural selection acts upon the average consequences of particular traits. This means that a particular trait may entail some costs, but as long as the cost/benefit analysis is favourable (i.e. on average it increases inclusive fitness over the individual's lifetime), the trait can be favoured. For example, in hedge sparrows (*Prunella modularis*), females often mate with multiple males, so a male cannot be sure which eggs he sired (Davies et al. 1992). As the males invest in parental care, they must estimate their paternity. This potentially difficult task is achieved via a simple approximation: the amount of uninterrupted access they had with the female (Burke et al. 1989). This rule is favoured despite the fact that about 15% of males estimate their paternity incorrectly and raise chicks of another male (Burke et al. 1989). This fitness cost is carried, as it seems the males are unable to achieve any better with the information they have. As we mentioned above, humans react to a subconscious cue of being watched by behaving more cooperatively. While the cues we respond to usually correlate with being watched, we do make mistakes. For example, if pictures of eyes, or stylized eye-like drawings are displayed, we will contribute more to public goods games, donate more to honesty boxes or litter less (Haley and Fessler 2005; Bateson et al. 2006; Burnham and Hare 2007; Ernest-Jones et al. 2011). Even seeing three dots configured as a down-pointing triangle (resembling two eyes and a mouth), instead of an up-pointing triangle, is enough to make males offer significantly more in a dictator game (Rigdon et al. 2009).

The lesson here is that if a few individuals in a population act irrationally, it does not automatically imply that their behaviour is maladaptive or that it needs a specific evolutionary explanation. We expect natural selection to minimize the average cost of errors, but we do not expect it to eliminate them. This may mean an odd individual makes a very costly error, or that everyone who performs the behaviour may incur occasional costs. Psychologists refer to this idea as 'error management theory' (Funder 1987; Haselton and Buss 2000; Haselton et al. 2005; McKay and Efferson 2010). As these errors in themselves are not adaptive, to seek an explanation for them in isolation is not possible.

Second, natural selection prefers cheap solutions. In response to the varied constraints organisms face, natural selection tends to favour traits of sundial-like design, by which we mean adaptations that are quick and cheap, but do the job, rather than are time-consuming, costly and precise. Biologists call these rules of thumb; psychologists and social scientists refer to them as heuristic biases (Kahneman and Tversky 1972; Tversky and Kahneman 1974; Gilovich et al. 2002; Gigerenzer and Gaissmaier 2011). For example, the parasitoid wasp *Trichogramma* uses a rule of thumb to estimate the volume of a sphere. This sphere is the egg of its host species. Its estimate must be accurate so that it lays the right number of its own eggs inside – too many and the larvae will run out of food, too few and it reduces its reproductive success. The wasp has reduced this complex calculation to a single proxy measure, the angle her head makes with her front leg when balanced on the egg (Wehner 1987). Since the 1970s, psychologists have discovered that humans rely on a diverse array of heuristic biases in many aspects of our decision making (Gilovich et al. 2002; Pohl 2004; Haselton et al. 2005). For example, we use heuristics to make estimates of relatedness, which allows us to avoid incest and to direct cooperative behaviours toward those we share genes with. Specifically, individuals are treated as closer relatives if there was a longer period of association during their childhood (Westermarck 1921; Lieberman et al. 2003; Corriveau and Harris 2009), if they bear a facial resemblance (DeBruine et al. 2008), if they speak the same dialect (Shutts et al. 2009) or if they smell familiar (Russel 1976; Russel et al. 1983; Porter and Cernoch 1983; Olsson et al. 2006).

Heuristics are cheap, but they are only adapted to maximize an aspect of fitness in a particular situation. They enable fast decision making and are reliable in commonly encountered situations (such as the example above of childhood association with close relatives), but they can cause us to behave in irrational ways, particularly when faced with novel situations. For this reason, the approximation used by the *Trichogramma* wasp is only accurate over the natural range of egg sizes she encounters; if presented with an abnormally small or large host egg, her rule of thumb fails and she would lay too many or too few eggs (Wehner 1987). In humans, the approximations for relatedness we use for incest avoidance cause problems in China and Taiwan where occasionally parents adopt a female infant and rear her with their son, whom she will eventually marry. Compared with marriages where spouses were raised apart, such marriages result in lower fertility and higher divorce rates (Wolf 1995). There is a vast and growing literature detailing how heuristic biases cause humans to make irrational decisions, particularly when faced with complex calculations or unfamiliar environments (Gilovich et al. 2002; Pohl 2004; Haselton et al. 2005).

Third, the solution to fitness trade-offs can change across time and circumstances. Behaviours may not appear to maximize anything, because different fitness components are traded off against each other and, as our circumstances and life stages change, these different components will change in importance (Stearns 1992). For example, we may be risk prone as adolescents, when our desire for social status outweighs our desire to avoid danger, and become risk averse once we have children, as our desire to keep our children and ourselves from harm outweighs any desire to be 'cool' (Steinberg 2008). Similarly, an old lung- or skin-cancer patient will probably wish they had not taken up smoking or used sunbeds when young. To an economist, such remorsefulness may appear irrational due to the inconsistency of preferences, but we are not designed to behave consistently, but to make the best of the situations we find ourselves in, continually altering our preferences throughout our lives in response to what gives us pleasure.

Fourth, novel environments produce non-optimal behaviour. Even behaviourally flexible organisms can only be expected to respond optimally within the range of natural variation they or their species have encountered before. Therefore in unfamiliar environments (such as a laboratory) organisms will make mistakes. In humans, this source of error is variously known as the 'mismatch hypothesis' (Hagen and Hammerstein 2006), artifact biases (Haselton et al. 2005) or, in cooperation research, the 'big mistake hypothesis' (Boyd and Richerson 2005b). As the physical and cultural environment humans live in is radically different from the one we evolved in, we may exhibit traits that are not adapted for our current environment. For example, in our ancestral environment, individuals with strong preferences for sugary, fatty and salty foods were favoured, as they sought out higher-quality diets. Today in developed countries, these evolved preferences are contributing to obesity and heart disease (Birch 1999). Similarly, strong sexual desires drove males to find the best mates and more willing mates. Now they also maintain a multibillion-dollar-a-year porn industry. It is not immediately clear whether suffering from heart disease or spending income on pornography is maladaptive or not, but this is not the point; and attempts to find out are often misguided. The point is that if these behaviours are not offset by other benefits then they will be selected against, providing the environment remains stable for a sufficiently long evolutionary time. For example, it is possible that the factors contributing to heart disease may have inclusive fitness benefits in early life, and that an attraction towards pornography may be a relatively harmless side effect of a beneficial sex drive. This idea is particularly relevant when studying cooperation because of the importance that demography and population structure have in the origin and maintenance of sociality.

Fifth, it is hard to measure fitness. Even when a behaviour is adaptive, the parameters being measured may be unable to detect the relationship. For example, behavioural ecologists use proxies of fitness, such as number of offspring, number of mates, longevity, food intake, body size or antler length. Better proxies (i.e. those more closely correlated with inclusive fitness) are expected to result in a better fit between the model prediction and the observed behaviours. Models that assume people are designed to maximize their income or social status suffer from the same problem as models that assume a stag is designed to maximize its antler length or a bee its nectar intake; they will only be accurate insofar as income, social status, antler length or nectar intake are accurate proxies for inclusive fitness. Furthermore (due to the four reasons highlighted above), behaviours are more likely to appear mistaken (i.e. the adaptive relationship will be missed) if observed for only a short time, in isolation, in a single individual, or when the individual is in an unfamiliar environment.

Sixth, and finally, not all traits are the target of selection. This means that the existence of such traits, which may be genetically or culturally inherited, cannot be understood with economic tools. For example, a genetic trait of interest may be a by-product of selection for another gene; this applies to fleece colour of Soay sheep (*Ovis aries*; Gratten et al. 2007) and red hair colour in humans (Valverde et al. 1995). In Soay sheep, the gene for dark fleeces is favoured as it is tightly linked (i.e. physically close on the chromosome) to a gene that is essential for cell protein function (Gratten et al. 2007). Similarly in humans, red hair has no adaptive value; however, the gene for red hair also causes pale skin, which is selected for when there is little sunshine, as it increases vitamin D synthesis from UVB radiation, which prevents rickets (Jablonski and Chaplin 2000). Adaptive explanations may help explain the fundamental desires that drive our culturally inherited behaviours. However, they say little about the cultural activities such as music, literature, sport, fashion and art that we undertake to fulfill those desires. In other words, much of what we value does not have an adaptive explanation so we do not believe that the minutiae of human behaviour can be studied using economic tools.

## 2.6 IS MEASURABLY OPTIMAL BEHAVIOUR EVER EXPECTED?

As optimal behaviour is not to be expected in the natural world, behavioural ecologists do not use economic tools to test whether animals behave optimally. Instead, they use them to make testable predictions that can help them understand why animals show the adaptations they do. Behavioural ecology is typically a *qualitative* science, where evolutionary explanations are established by comparative studies across different conditions or between species (Davies et al. 2012). Having said that, in the natural world, there are very rare cases where optimality models do accurately predict *quantitative* differences. The best-known case of such 'as-if' optimization is that of sex ratios, where very simple models can make accurate quantitative predictions (Charnov 1982; West 2009). For example, there is a tight fit between observed sex ratios in the fig-pollinating wasps and those predicted by theory, which says that as the number of females laying eggs in a fruit increases, the sex ratios in the broods should become less female biased (West et al. 2000).

Why do the models and data match so well in the case of sex ratios? The first reason is that sex allocation represents a very basic trade-off (male vs female), so it's easy to construct models that accurately reflect the constraints and choices available to the organism. Second, the traits can be measured precisely, as it is relatively easy to sex offspring and count the number of surviving grandchildren. Third, the sex ratio is very strongly tied to fitness, so it is strongly selected for, meaning that natural selection will be the dominant evolutionary force, allowing the trait to get close to the goal of optimality. Finally, the sex ratio is a neat quantitative trait, with the expectation that mutations of small effect can readily arise that will nudge the sex ratio by a small amount, allowing a great precision of adaptation even in a very simple organism such as the fig wasp. The example of sex-ratio research indicates when as-if optimality is most likely to be observed: when there is a simple set of choices, where decisions can be precisely quantified, when there are high stakes and where the individual can control their choice. However it should be noted that even in the case of sex ratios, deviations from predictions are observed, but these deviations are not random. In fact, as predicted, they negatively correlate with the strength of selection and with the reliability of environmental cues (West et al. 2002). Other instances of as-if optimization in animals are found in clutch sizes and various aspects of foraging behaviour (Davies et al. 2012).

The human capacity for learning, reasoning and behavioural flexibility means that, despite the mistakes we sometimes make, in a wide range of cases we do behave in ways that match the predictions of optimality models. Indeed, we are so good at fine-tuning our real-time behaviour toward achieving specific goals that it is typically studied using rational choice models. The maximand for a rational choice model is an individual's utility. Utility is maximized over short timescales and correlates to desires or wants (Marshall 1920) so can be related to inclusive fitness (Grafen 1998). It is a numerical representation of a preference ranking over a set of alternatives. Given money's properties as a universal exchange rate, it is no surprise that in today's world so many people have decided that making money is an effective way to be happy. This justifies the oft-made simplifying assumption that utility represents a single preference for income maximization.

So long as we carefully model the constraints and choices humans face, rational choice models can accurately describe human behaviour, i.e. we do achieve as-if optimization of utility in real time in many situations (Tversky and Kahneman 1986). Learning from the example of as-if optimality in nature, we predict that human behaviour is more likely to quantitatively match the predictions of rational choice models when the decision can be precisely measured (as is the case for market behaviour), the choices are simple or routine (so there is a clear optimal decision and we do not need to rely on heuristics), the stakes are high (so the individual is motivated to 'care' about the outcome), and the individual is in total control of their choice. It is often hard to give economic explanations for our actions, particularly when they affect others or have other far-reaching consequences. When attempting to model such behaviour, care is needed to avoid unrealistic assumptions, for example by only specifying cognitively feasible concerns. Furthermore, parsimonious explanations are often possible with simpler models that only focus on personal costs and benefits. For example, an experimental participant that acts in a way that benefits or harms others may not be motivated beyond personal concerns; such externalities may be a by-product of experimental design (Kümmerli et al. 2010).

Humans are so capable of achieving as-if optimization of utility in the way rational choice models predict, it is no surprise that most economic theories assume we behave in this way. However, we do not match the predictions of rational choice models all the time (Sen 1977). This is especially true for non-market behaviour. In the previous section, we discussed why our behaviour will often not appear to maximize anything; economists will be familiar with many of these reasons. In addition to these reasons, however, the predictions about human behaviour arising from the expectation that individuals will strive to maximize their inclusive fitness will not always match the predictions of models which assume we strive for real-time utility maximization. This is because income and happiness are similar to body mass and antler length, in that they are only proxies for inclusive fitness, and actual inclusive fitness is the only maximand that the average lifetime consequences of a person's actions might be consistent with. Furthermore humans are not expected to have consistent preferences during their lives (as rational choice theory assumes). Instead, preferences vary in importance and will shift over time as an individual's personal circumstances and knowledge changes. These preference changes may seem irrational from an economic perspective, but they can make sense from an inclusive fitness point of view. Most importantly, whether we appear to solve a particular problem in an economically rational way will depend on the time frame of the problem. Over short time frames, we do maximize our pleasure. However, we are not designed to maximize pleasure in the same way over our lifetime, as what gives us pleasure will continually change. Finally, optimization models, no matter how sophisticated or complex, cannot describe all our behaviour as we are a product of evolution, not just natural selection. In other words, we may be good, but we are not optimally designed.

## 2.7 CONCLUSION

How would an alien biologist that was capable of observing our behaviours and reading our minds sum up what humans maximize? They would probably say that, intentionally, humans do not aim to maximize any one thing over their lifetimes, but do generally try to maximize their pleasure and minimize their pain over short time frames. The alien may note that we receive pleasure and pain from adaptively sensible sources, and that we are very good at learning or inventing new ways to increase our pleasure. They may also note that the happiest people are those whose lives have improved constantly, rather than those who started at the top and just had to stay there. Finally, while realizing that humans engage in many instances of cooperation, their energies appear to be mostly directed towards benefiting themselves, their reproductive opportunities, and the wellbeing of their relatives. While finding them fascinating for many reasons, the alien would conclude that humans, along with all other organisms, are best described as striving to maximize their inclusive fitness over their lifetimes, yet in imperfect and non-optimal, but often predictable, ways.

## REFERENCES

Axelrod, R. and Hamilton, W. D. 1981. 'The evolution of cooperation', *Science* 211: 1390–1396.

Bateson, M., Nettle, D. and Roberts, G. 2006. 'Cues of being watched enhance cooperation in a real-world setting', *Biology Letters* 2: 412–414.

Baum, W. B. 1994. *Understanding behaviourism: science, behaviour and culture.* New York: HarperCollins.

Bergmüller, R., Bshary, R., Johnstone, R. A. and Russell, A. F. 2007. 'Integrating cooperative breeding and cooperation theory', *Behavioural Processes* 76: 61–72.

Birch, L. 1999. 'Development of food preferences', *Annual Review of Nutrition* 19: 41–62.

Boyd, R. and Richerson, P. J. 1985. *Culture and the evolutionary process.* Chicago University Press.

2005a. *The origin and evolution of cultures.* Oxford University Press.

2005b. 'Solving the puzzle of human cooperation', in: Levinson, S. (ed.), *Evolution and culture*. Cambridge, MA: MIT Press.

1987. 'An anthropological perspective on obesity', *Annals of the New York Academy of Sciences* 499: 29–46.

Burke, T., Davies, N. B., Bruford, M. W. and Hatchwell, B. J. 1989. 'Parental care and mating behaviour of polyandrous dunnocks *Prunella modularis* related to paternity by DNA fingerprinting', *Nature* 338: 249–251.

Burnham, T. C. and Hare, B. 2007. 'Engineering human cooperation: does involuntary neural activation increase Public Goods contributions?' *Human Nature* 18: 88–108.

Charnov, E. L. 1982. *The theory of sex allocation*. Princeton University Press.

Clutton-Brock, T. H. and Parker, G. A. 1995. 'Punishment in animal societies', *Nature* 373: 209–216.

Cohen, D., Nisbett, R. E., Bowdle, B. F. and Schwarz, N. 1996. 'Insult, aggression, and the southern culture of honor: an "experimental ethnography"', *Journal of Personality and Social Psychology* 70(5): 945–960.

Corriveau, K. H. and Harris, P. L. 2009. 'Choosing your informant: weighing familiarity and past accuracy', *Developmental Science* 12: 426–437.

Darwin, C. R. 1859. *The origin of species*. London: John Murray.

1871. *The descent of man and selection in relation to sex*. London: John Murray.

Davies, N. B., Hatchwell, B. J., Robson, T. and Burke, T. 1992. 'Paternity and parental effort in dunnocks *Prunella modularis*: how good are male chick-feeding rules?' *Animal Behaviour* 43: 729–745.

Davies, N. B., Krebs, J. R. and West, S. A. 2012. *An introduction to behavioural ecology*, 4th edn. Hoboken, NJ: Wiley-Blackwell.

Dawkins, R. 1976. *The selfish gene*. New York: Oxford University Press.

Dawson, A. 2008. 'Control of the annual cycle in birds: endocrine constraints and plasticity in response to ecological variability', *Philosophical Transactions of the Royal Society B* 363: 1621–1633.

DeBruine, L., Jones, B. C., Little, A. and Perrett, D. I. 2008. 'Social perception of facial resemblance in humans', *Archives of Sexual Behaviour* 37: 64–77.

de Quervain, D. J. F., Fischbacher, U., Treyer, V., Schellhammer, M., Schnyder, U., Buck, A. and Fehr, E. 2004. 'The neural basis of altruistic punishment', *Science* 305: 1254–1258.

Ehlinger, T. J. 1989. 'Learning and individual variation in bluegill foraging: habitat specific technique', *Animal Behaviour* 38: 643–658.

El Mouden, C., West, S. A. and Gardner, A. 2010. 'The enforcement of cooperation by policing', *Evolution* 64: 2139–2152.

Ernest-Jones, M., Nettle, D. and Bateson, M. 2011. 'Effects of eye images on everyday cooperative behaviour: a field experiment', *Evolution and Human Behaviour* 32: 172–178.

Fehr, E. and Fischbacher, U. 2003. 'The nature of human altruism', *Nature* 425: 785–791.

Fehr, E. and Gächter, S. 2002. 'Altruistic punishment in humans', *Nature* 415: 137–140.

Fisher, R. A. 1930 *The genetical theory of natural selection*. Oxford: Clarendon.

Fiske, A. P. 1999. 'Learning culture the way informants do: observation, imitation, and participation'. Unpublished manuscript.

Frank, S. A. 1995. 'Mutual policing and repression of competition in the evolution of cooperative groups', *Nature* 377: 520–522.

2003. 'Repression of competition and the evolution of cooperation', *Evolution* 57: 693–705.

Funder, D. C. 1987. 'Errors and mistakes. Evaluating the accuracy of social judgement', *Psychological Bulletin* 101: 75–90.

Gangestad, S. W., Thornhill, R. and Garver, C. 2002. 'Changes in women's sexual interests and their partners' mate retention tactics across the menstrual cycle: evidence for shifting conflicts of interest', *Proceedings of the Royal Society B* 269: 975–982.

Gangestad, S. W., Thornhill, R. and Garver-Apgar, C. E., 2005. 'Adaptations to ovulation', in: Buss, D. M. (ed.), *The handbook of evolutionary psychology*. Hoboken, NJ: John Wiley & Sons.

Gardner, A. 2009. 'Adaptation as organism design', *Biology Letters* 5: 861–864.

Gardner, A. and West, S. A. 2004. 'Cooperation and punishment, especially in humans', *American Naturalist* 164: 753–764.

Gigerenzer, G. and Gaissmaier, W. 2011. 'Heuristic decision making', *Annual Review of Psychology* 62: 451–8

Gilovich, T., Griffin, D. and Kahneman, D. 2002. *Heuristics and biases: the psychology of intuitive judgment*. Cambridge University Press.

Gintis, H., Bowles, S., Boyd, R. and Fehr, E. 2005. *Moral sentiments and material interests – the foundations of cooperation in economic life*. Cambridge, MA: MIT Press.

Grafen, A. 1998. 'Fertility and labour supply in *Femina economica*', *Journal of Theoretical Biology* 194: 429–455.

2002. 'A first formal link between the Price equation and an optimization program', *Journal of Theoretical Biology* 217: 75–91.

2006. 'Optimization of inclusive fitness', *Journal of Theoretical Biology* 238: 541–563.

2007. 'The formal Darwinism project: a mid-term report', *Journal of Evolutionary Biology* 20: 1243–1254.

Gratten, J., Beraldi, D., Lowder, B. V., McRae, A. F., Visscher, P. M., Pemberton, J. M. and Slate, J. 2007. 'Compelling evidence that a single nucleotide substitution in TYRP1 is responsible for coat-colour polymorphism in a free-living population of Soay sheep', *Proceedings of the Royal Society B* 274: 619–626.

Hagen, E. H. and Hammerstein, P. 2006. 'Game theory and human evolution: a critique of some recent interpretations of experimental games', *Theoretical Population Biology* 63: 339–348.

Halama, K. J. and Reznick, D. N. 2001. 'Adaptation, optimality, and the meaning of phenotypic variation in natural populations', in: Orzack, S. H. and Sober, E. (eds.) *Adaptationism and Optimality*. Cambridge University Press.

Haley, K. J. and Fessler, D. M. T. 2005. 'Nobody's watching? Subtle cues affect generosity in an anonymous economic game', *Evolution and Human Behavior* 26: 245–256.

Hamilton, W. D. 1964. 'The genetical evolution of social behaviour, I and II', *Journal of Theoretical Biology* 7: 1–52.

1970. 'Selfish and spiteful behaviour in an evolutionary model', *Nature* 228: 1218–1220.

1996. *Narrow roads of gene land*, vol. I: *Evolution of social behaviour*. Oxford: W. H. Freeman.

Harbaugh, W. T., Mayr, U. and Burghart, D. R. 2007. 'Neural responses to taxation and voluntary giving reveal motives for charitable donations', *Science* 316: 1622–1625.

Hardin, G. 1968. 'The tragedy of the commons', *Science* 162: 1243–1248.

Haselton, M. G. and Buss, David. 2000. 'Error management theory: a new perspective on biases in cross-sex mind reading', *Journal of Personality and Social Psychology* 78: 81–91.

Haselton, M. G. and Gangestad, S. W. 2006. 'Conditional expression of women's desires and men's mate guarding across the ovulatory cycle', *Hormones and Behaviour* 49: 509–551.

Haselton, M. G., Mortezaie, M., Pillsworth, E. G., Bleske-Recheck, A. E. and Frederick, D. A. 2007. 'Ovulation and human female ornamentation: near ovulation, women dress to impress', *Hormones and Behavior* 51: 40–45.

Haselton, M. G., Nettle, D. and Andrews, P. W. 2005. 'The evolution of cognitive bias', in: Buss, D. M. (ed.), *Handbook of evolutionary psychology*. Hoboken, NJ: Wiley.

Henrich, J. 2005. 'Cultural evolution of human cooperation', in: Boyd, R. and Richerson, P. (eds.), *The origin and evolution of cultures*. Oxford University Press.

2008. 'A cultural species', in: Brown, M. (ed.), *Explaining culture scientifically*. Seattle, WA: University of Washington Press

Henrich, J. and McElreath, R. 2003. 'The evolution of cultural evolution', *Evolutionary Anthropology* 12(3): 123–135.

Herre, E. A. 1985. 'Sex ratio adjustment in fig wasps', *Science* 228: 896–898.

1987. 'Optimality, plasticity and selective regime in fig wasp sex ratios', *Nature* 329: 627–629.

Heyes, C. M. and Galef, B. G. 1996. *Social learning in animals: the roots of culture*. San Diego Academic Press.

Hollis, K. L., Pharr, V. L., Dumas, M. J., Britton, G. B. and Field, J. 1997. 'Classical conditioning provides paternity advantage for territorial male blue gouramis (*Trichogaster trichopterus*)', *Journal of Comparative Psychology* 111: 219–225.

Jablonski, N. G. and Chaplin, C. 2000. 'The evolution of human skin coloration', *Journal of Human Evolution* 39: 57–106.

Kahneman, D. and Tversky, A. 1972. 'Subjective probability: a judgment of representativeness', *Cognitive Psychology* 3: 430–454

Kiers, E. T., Rousseau, R. A., West, S. A. and Denison, R. F. 2003. 'Host sanctions and the legume–rhizobium mutualism', *Nature* 425: 78–81.

Kokko, H., Johnstone, R. A. and Clutton-Brock, T. H. 2001. 'The evolution of cooperative breeding through group augmentation', *Proceedings of the Royal Society B* 268: 187–196.

Kümmerli, R., Burton-Chellew, M. N., Ross-Gillespie, A. and West, S. A. 2010. 'Resistance to extreme strategies, rather than prosocial preferences, can explain human cooperation in public goods games', *Proceedings of the National Academy of Science, USA* 107: 10125–10130.

Lasseka, W. D. and Gaulin, S. J. C. 2008. 'Waist–hip ratio and cognitive ability: is gluteofemoral fat a privileged store of neurodevelopmental resources?', *Evolution and Human Behavior* 29: 26–34.

Law, R. 1979. 'Optimal life histories under age-specific predation', *American Naturalist* 114: 399–417.

Lehmann, L. and Keller, L. 2006. 'The evolution of cooperation and altruism: a general framework and classification of models', *Journal of Evolutionary Biology* 19: 1365–1378.

Lehmann, L., Rousset, F., Roze, D. and Keller, L. 2007. 'Strong-reciprocity or strong ferocity? A population genetic view of the evolution of altruistic punishment', *American Naturalist* 170: 21–36.

Leigh, E. G. 1971. *Adaptation and diversity.* San Francisco, Freeman, Cooper & Co.

Lieberman, D., Tooby, J. and Cosmides, L. 2003. 'Does morality have a biological basis? An empirical test of the factors governing moral sentiments relating to incest', *Proceedings of the Royal Society B* 270: 819–826.

Mahometa, M. J. and Domjan, M. 2005. 'Classical conditioning increases reproductive success in Japanese quail, *Coturnix japonica*', *Animal Behaviour* 69: 983–989.

Marshall, A. 1920. *Principles of economics: an introductory volume*, 8th edn. London: Macmillan.

Maynard Smith, J. 1982. *Evolution and the theory of games.* Cambridge University Press.

McGrew, W. C. 2004. *The cultured chimpanzee: reflections on cultural primatology.* Cambridge University Press.

McKay, R. and Efferson, C. 2010. 'The subtleties of error management', *Evolution and Human Behavior* 31: 309–319.

Miller, G. F., Tybur, J. and Jordan, B. 2007. 'Ovulatory cycle effects on tip earnings by lap-dancers: economic evidence for human estrus?', *Evolution and Human Behavior* 28: 375–381.

Moll, J., Krueger, F., Zahn, R., Pardini, M., de Oliviera-Souza, R. and Grafman, J. 2006. 'Human fronto–mesolimbic networks guide decisions about charitable donation', *Proceedings of the National Academy of Science, USA* 103: 15623–15628.

Møller, A. P. and Jennions, M. D. 2002. 'How much variance can be explained by ecologists and evolutionary biologists?', *Oecologia* 132: 492–500.

Olsson, S. B., Barnard, J. and Turri, L. 2006. 'Olfaction and identification of unrelated individuals: examination of the mysteries of human odor recognition', *Journal of Chemical Ecology* 32: 1635–1645.

Paley, W. 1802. *Natural theology*. London: Wilks & Taylor.

Parker, G. A. and Maynard Smith, J. 1990. 'Optimality theory in evolutionary biology', *Nature* 348: 27–33.

Pearce, D. 1997. *Animal learning and cognition: an introduction*, 2nd edn. Hove: Psychology Press.

Penton-Voak, I. S., Perrett, D. I., Castles D. L., Kobayashi, T., Burt, D. M., Murray, L. K. and Minamisawa, R. 1999. 'Menstrual cycle alters face preference', *Nature* 399: 741–742.

Perrins, C. M. 1979. *British tits*. London: HarperCollins.

Pettit, P. 2005. 'The rise of modern humans', in: Scarre, C. (ed). *The human past*. London: Thames & Hudson.

Pohl, R. 2004. *Cognitive illusions: a handbook on fallacies and biases in thinking, judgement and memory*. Hove: Psychology Press.

Porter, R. H. and Cernoch, J. M. 1983. 'Maternal recognition of neonates through olfactory cues', *Physiology and Behavior* 30: 151–154.

Quartz, S. R. 2002. *Liars, lovers and heroes*. New York: William Morrow.

Quartz, S. and Sejnowski, T. 2000. 'Constraining constructivism: cortical and sub-cortical constraints on learning in development', *Behavioral and Brain Sciences* 23(5): 785–792.

Richerson, P. J. and Boyd R. 2004. *Not by genes alone*. University of Chicago Press.

Rigdon, M., Ishii, K., Watabe, M. and Kitayama, S. 2009. 'Minimal social cues in the dictator game', *Journal of Economic Psychology* 30: 358–367.

Rilling, J. K., Gutman, D. A., Zeh, T. R., Pagnoni, G., Berns, G. S. and Kilts, D. S. 2002. 'A neural basis for social cooperation', *Neuron* 35: 395–402.

Roberts, S. D., Havlicek, J., Flegr, J., Hruskova, M., Little, A. C., Jones, B. C., Perrett, D. I. and Petrie, M. 2004. 'Female facial attractiveness increases during the fertile phase of the menstrual cycle', *Proceedings of the Royal Society B* 271: S270–S272.

Russel, M. J. 1976. 'Human olfactory communication', *Nature* 260: 520–522.

Russel, M. J., Mendelson, T. and Peeke, H. V. S. 1983. 'Mothers' identification of their infant's odor', *Ethology and Sociobiology* 4: 29–31.

Sachs, J. L., Mueller, U. G., Wilcox, T. P. and Bull, J. J. 2004. 'The evolution of cooperation', *Quarterly Review of Biology* 79: 135–160.

Segall, M., Campbell, D. and Herskovits, M. J. 1966. *The influence of culture on visual perception*. New York: Bobbs-Merrill Co.

Sen, A. K. 1977. 'Rational fools: a critique of the behavioural foundations of economic theory', *Philosophy & Public Affairs* 6(4): 317–344.

Shutts, K., Kinzler, K. D., McKee, C. B., and Spelke, E. S. 2009. 'Social information guides infants' selection of foods', *Journal of Cognition and Development* 10: 1–17.

Simões-Lopes, P. C., Fabián, M. E. and Menegheti, J. O. 1998. 'Dolphin interactions with the mullet artisanal fishing on southern Brazil: a qualitative and quantitative approach', *Revista Brasileira de Zoologia* 15(3): 709–726.

Simon, H. A. 1990. 'A mechanism for social selection and successful altruism', *Science* 250: 1665–1668.

Singh, D. 1993. 'Adaptive significance of female physical attractiveness: role of waist–hip ratio', *Journal of Personality and Social Psychology* 65: 293–307.

Smith, T. B. and Skúlason, S. 1996. 'Evolutionary significance of resource polymorphisms in fish, amphibians and birds', *Annual Review of Ecology and Systematics* 27: 111–113.

Stanfey, A. G. 2007. 'Social decision-making: insights from game theory and neuroscience', *Science* 318: 598–602.

Stearns, S. 1992. *The evolution of life histories*. Oxford University Press.

Steinberg, L. 2008. 'A social neuroscience perspective on adolescent risk-taking', *Development Review* 28: 78–106.

Sugiyama, L. S. 2004. 'Is beauty in the context-sensitive adaptations of the beholder?', *Evolution and Human Behaviour* 25: 51–62.

Sullivan, K. A. 1988. 'Age-specific profitability and prey choice', *Animal Behaviour* 36: 613–615.

Swami, V. and Tovee, M. J. 2006. 'Does hunger influence judgments of female physical attractiveness?' *British Journal of Psychology* 97: 353–363.

Taborsky, M. 1994. 'Sneakers, satellites and helpers: parasitic and cooperative behavior in fish reproduction', *Advances in the Study of Behaviour* 21: 1–100.

Tooby, J. and Cosmides, L. 1990. 'The past explains the present: emotional adaptations and the structure of ancestral environments', *Ethnology and Sociobiology* 11(4–5): 375–424.

Tomasello, M. 1999. *The cultural origins of human cognition*. Cambridge, MA: Harvard University Press.

Trivers, R. L. 1971. 'The evolution of reciprocal altruism', *The Quarterly Review of Biology* 46(1): 35–57.

Tversky, A. and Kahneman, D. 1974. 'Judgment under uncertainty: heuristics and biases', *Science* 185: 1124–1131.

    1986. 'Rational choice and the framing of decisions', *Journal of Business* 59(4): S251–S278.

Valverde, P., Healy, E., Jackson, I., Rees, J. L. and Thody, A. J. 1995. 'Variants of the melanocyte-stimulating hormone receptor gene are associated with red hair and fair skin', *Nature Genetics* 11: 328–330.

Warren, J. M. 1965. 'Primate learning in comparative perspective', in: Schrier, A. M., Harlow, H. F. and Stollnitz, F. (eds.) *Behavior of nonhuman primates: modern research trends*. New York: Academic Press.

Wehner, R. 1987. '"Matched filters" – neural models of the external world', *Journal of Comparative Physiology A* 161: 511–531.

West, S. A. 2009. *Sex allocation*. Princeton University Press.

West, S. A., El Mouden, C. and Gardner, A. 2011. 'Sixteen common misconceptions about the evolution of cooperation in humans', *Evolution and Human Behavior* 32: 231–262.

West, S. A., Griffin, A. S. and Gardner, A. 2007a. 'Social semantics: how useful has group selection been?' *Journal of Evolutionary Biology* 21: 374–385.

2007b. 'Evolutionary explanations for cooperation', *Current Biology* 17: R661–R672.

West, S. A., Herre, E. A. and Sheldon, B. C. 2000. 'The benefits of allocating sex', *Science* 290: 288–290.

West, S. A., Reece, S. E. and Sheldon, B. C. 2002. 'Sex ratios', *Heredity* 88: 117–124.

Westermarck, E. A. 1921. *The history of human marriage*, 5th edn. London: Macmillan

Whiten, A. and Ham, R. 1992. 'On the nature and evolution of imitation in the animal kingdom: a reappraisal of a century of research', in: Slater, P. J. B, Rosenblatt, J. S., Beer, C. and Milkinski, M. (eds.), *Advances in the study of behaviour*, 21. New York: Academic Press.

Williams, G. C. 1966. *Adaptation and natural selection*. Princeton University Press.

Wolf, A. P. 1995. *Sexual attraction and childhood association: a Chinese brief for Edward Westermarck*. Stanford University Press.

# CHAPTER 3

# Natural selection and rational decisions

*Alasdair I. Houston*

## 3.1 INTRODUCTION

Natural selection is often regarded as an optimizing agent, producing organisms that maximize reproductive success. A consequence is that natural selection results in organisms that meet the standards that are required for what is known as rational behaviour (Cooper 2001; Houston et al. 2007a). The view that rational behaviour amounts to meeting certain formal requirements is commonly adopted in economics and hence is called E-rationality by Kacelnik (2006). This approach does not pass judgement on the currency that is maximized. In contrast, a biological approach to rationality argues that the currency is given by the action of natural selection. This approach is what Kacelnik calls B-rationality.

There are two standard requirements for E-rationality in a situation of individual decision making under certainty, independence from irrelevant alternatives (IIA) and transitivity. IIA means that relative preference for one option over another is unchanged when options are added to or removed from the choice set (Luce 1959). In this chapter I concentrate on transitivity.

Transitivity means that if a decision maker prefers option *a* to option *b* and also prefers option *b* to option *c* then they will prefer option *a* to option *c*. It has often been claimed that the behaviour of humans (e.g. Tversky 1969; Tversky and Simonson 1993; Linares 2009; Day and Loomes 2010; Kalenscher et al. 2010) and other animals (e.g. honeybees (Shafir 1994) and gray jays (Waite 2001)) fails to satisfy this condition. In this chapter, I do not consider the problem of establishing that transitivity has been violated (Birnbaum 2010; Browne et al. 2010; Regenwetter et al. 2011). Instead I focus on the relationship between transitivity, rationality and natural selection. Violations of transitivity pose fundamental problems for theories of decision making because they suggest that there is no single scale that can be used to assign value to choices.

In introducing the concept of transitivity I tacitly assumed that when given two particular options, an agent will always choose the same option. This is not the case; animals' choices are typically stochastic rather than deterministic (Stephens and Krebs 1986; Rieskamp 2008). As a result, it is necessary to work with a stochastic form of preference. I use the notation $p(a, b)$ to represent the probability that option $a$ is chosen when the agent is presented with options $a$ and $b$. In the context of stochastic choice, there are various forms of transitivity (Luce and Suppes 1965; Fishburn 1973). The form that has been used in several of the papers that I discuss is strong stochastic transitivity (SST). It is defined as follows.

$$p(a,b) \geq 0.5 \text{ and } p(b,c) \geq 0.5$$

imply

$$p(a,c) \geq \max[p(a,b), p(b,c)].$$

SST is equivalent to the concept of substitutability (Tversky and Russo 1969). Choice satisfies substitutability if:

$$p(a,c) > p(b,c) \text{ implies } p(a,b) > 0.5$$

and

$$p(a,c) = p(b,c) \text{ implies } p(a,b) = 0.5.$$

SST is also equivalent to simple scalability (Tversky and Russo 1969). Choice satisfies simple scalability if there exist real-valued functions $\varphi$ and $u$ such that for any two options $a$, $b$: $p(a,b) = \varphi(u(a),u(b))$, where $\varphi$ is strictly increasing in its first argument and strictly decreasing in its second argument.

Theories of choice have been dominated by the idea that we can attach a value to each alternative in such a way as to determine the probability that it, rather than another alternative, is chosen. In particular, given a set of alternatives $a$, $b$, $c$ ... it is often assumed that we can assign values $u(a)$, $u(b)$, $u(c)$ ... such that the probability of choosing $a$ when given a choice between $a$ and $b$ is

$$p(a,b) = \frac{u(a)}{u(a) + u(b)} \tag{3.1}$$

Luce and Suppes (1965) refer to this as a 'strict utility' model. If Equation (3.1) holds, then choice satisfies simple scalability.

If options are ranked along a single dimension then choice will be transitive. Foraging options can often be characterized in terms of two dimensions, as the following examples show.

### 3.1.1 Energy and handling time

In the standard model of prey choice (Stephens and Krebs 1986), a forager encounters different types of prey. Each item of type $i$ yields an amount of energy $e_i$ and requires a time $h_i$ (the handling time) to capture and consume it. If the forager's aim is to maximize its rate of energetic gain then it should accept an item if and only if $e_i/h_i > \gamma$, where $\gamma$ is the maximum possible rate of energetic gain. Other currencies based on time and energy are discussed by McNamara and Houston (1997).

### 3.1.2 Mean and variance

Under rate maximization, only an option's mean energy and mean handling time are important. Rate maximization is an appropriate basis for decisions if each unit of energy produces the same increase in reproductive success. If this is not the case, then the means are no longer sufficient to determine optimal decisions. This can be clearly seen when choices are made at regular times that do not depend on previous choices. (In particular the handling time associated with an option does not influence when the next choice can be made.) In this case the value of an option depends not just on the resulting mean amount of energy but also on its distribution, often characterized by the variance. There is an extensive body of theory that derives optimal decisions from an organism's ecological circumstances (McNamara and Houston 1992; Houston and McNamara 1999). The subject area is known as risk sensitive foraging. (Risk in this context refers to variability; risk as danger is considered in the next section.)

### 3.1.3 Energy and predation risk

A foraging animal may only be able to increase its energy gain by also increasing its risk of being killed by a predator (Lima and Dill 1990; Lima 1998; Bednekoff 2007). In other words there is a trade-off between energy and predation. The optimal decision when faced with such a trade-off can be found by using reproductive value as a common currency (McNamara and Houston 1986). Reproductive value quantifies how the ability to leave descendants in the distant future depends on an organism's state (Houston and McNamara 1999). The state of a foraging animal is often taken to be its energy reserves. Let $V(x)$ be the reproductive value as a function of energy reserves $x$. The benefit from foraging is an increase in $V$ as a result of an increase in $x$. The cost is the lost reproductive value if the forager is killed. Houston and McNamara (1989) show that the optimal decision maximizes $\gamma V' - \mu V$, where $\gamma$ is the rate of energetic gain and $\mu$ is the rate of predation. The term $\gamma V'$ is the rate of increase in reproductive value as a result of foraging and the term $\mu V$ is the rate of decrease in reproductive value because of predation, so the whole expression is the net rate of increase in reproductive value.

Several studies of intransitivity have been based on choosing between foraging options, e.g. Shafir (1994), Waite (2001) and Latty and Beekman (2011). Choice between alternatives that differ on more than one dimension has also been investigated in the context of an animal choosing a mate (Bateson and Healy 2005; Royle et al. 2008; Reaney 2009).

Birnbaum and LaCroix (2008) point out that when options differ on more than one dimension, ways of making a decision can be divided into two categories.

### Integrative approaches

In these approaches, values on different dimensions are combined to give a single overall value for an option. This is the basis of normative models of decision making. For example, when foraging options differ in terms of energy and handling time, they can be evaluated by computing $e - \gamma h$ (McNamara 1982). When they differ in terms of intake rate and rate of mortality they can be compared in terms of the net rate of increase in reproductive value (Houston and McNamara 1989). Because the single value assigned to an option can depend on the other available options (e.g. Loomes and Sugden 1982; Houston 1991), this approach does not necessarily result in transitive decisions.

### One dimension at a time

Models in this class do not combine values from the different dimensions. Instead, each dimension is considered in isolation ('one-reason decision making', Katsikopoulos and Gigerenzer 2008). Because dimensions are considered one at a time, there is no explicit trade-off between values on different dimensions (cf. Brandstatter et al. 2006). As a result, one dimension at a time (ODAT) procedures can perform badly in some circumstances. This should not be seen as a reason to dismiss such rules. The idea of bounded or ecological rationality (Simon 1956, 1990; Todd and Gigerenzer 2003, 2007) is that simple and successful heuristics exploit regularities in the environment. Given certain relationships between possible options, ODAT procedures can perform well. As an example, assume that a forager has to choose between options that differ in energy $e$ and handling time $h$. If large food items have high $e$ and $h$, then an optimal choice requires knowledge of the trade-off between $e$ and $h$. This trade-off is captured by $\gamma$, which converts time spent on an option to a potential amount of energy that is lost by choosing the option. Now imagine an environment in which all options have the same value of $h$. In such an environment there is no conflict between selecting a high value of $e$ and a low value of $h$, and the optimal decision is given by choosing the option with the highest value of $e$. Although this is an extreme example, it can be argued that there are many circumstances in which good choices can be based on a single dimension.

I now consider the relationship between natural selection and intransitive decisions. In the first sort of explanation, selection acts on a decision rule and intransitivity is an unselected consequence. In the second sort of explanation, selection produces intransitivity as part of an optimal solution.

## 3.2 INTRANSITIVITY AS AN UNSELECTED CONSEQUENCE

One way to explain irrational behaviour in terms of natural selection is to make use of the fact that natural selection is not expected to produce organisms that compute the optimal solutions to every condition that they encounter (Hutchinson and Gigerenzer 2005; Houston 2009; McNamara and Houston 2009). It is widely argued that animals use rules that perform well but are not exactly optimal in all circumstances. Such rules may violate the requirements of rational choice in some circumstances. For example, Gigerenzer (2000) describes a rule that he calls minimalist. When faced with options that differ on more than one dimension, an agent using the minimalist rule picks a dimension at random and then selects the option that ranks highest on the chosen dimension. This rule can result in intransitive choices but it performs well compared with rules that are more complex (some of which always produce transitive decisions). Selecting a cue at random might not seem realistic, but violations of transitivity can occur when cues are used in a systematic order. In the procedure that Gigerenzer calls 'Take the Best', each option has a value on each of several cues. For each cue, the possible values are 'present' (+), 'absent' (−) or 'don't know'. When a decision has to be made between two options, they are first compared on cue 1. If one option has a + and the other has a −, then the one with the + is chosen. If cue 1 does not result in a decision, the procedure attempts to make a decision on the basis of cue 2, and so on until a decision is achieved.

As another example of an explanation based on rules that are not always optimal, Pavlic and Passino (2010) consider why animals in laboratory experiments fail to maximize rate of gain when given a choice between two options that differ in terms of reward magnitude and delay (essentially $e$ and $h$ in the prey-choice notation). When offered a choice between such options, animals may prefer a small reward after a short delay (SS) over a large reward after a long delay, even though SS options should not be taken by a forager that is maximizing its rate of gain. Pavlic and Passino claim that a preference for these options can be understood as a side effect of a rule that performs well outside the lab. They suggest that the rule estimates the rate $\gamma$ that an environment provides, and accepts an option if the associated value of $e/h$ is greater than $\gamma$ (cf. McNamara and Houston 1985). This rule learns the optimal diet when options are encountered sequentially. Pavlic and Passino argue that the typical lab procedure of depriving an animal of food before an experiment gives the animal a low initial estimate of $\gamma$. They assume that when offered a simultaneous choice of options, the animal first attends to the option with the lower value of $h$. If its value of $e/h$ is greater than $\gamma$, then the option is accepted. If not, then the other option is considered. If $\gamma$ is low enough, the animal will accept the SS items and will never learn about other items that have higher values of $e/h$. Although Pavlic and Passino see this as providing an explanation for the laboratory findings, it ignores the fact that most experimental procedures ensure that the subjects experience both options by having 'forced choice' (or 'no choice') trials as well as the trials on which the animal can choose (e.g. Rachlin and Green 1972; Green and Estle 2003). On forced choice trials only one option is available, and each of the two options is available on the same number of forced trials.

Another possible explanation for apparently irrational behaviour is based on differences in state (Schuck-Paim et al. 2004). I have pointed out that reproductive value $V$ provides a common currency for comparing options. In general, $V$ will depend on the animal's state, and so it is optimal for animals to make state-dependent decisions. For example, if state is the level of energy reserves, $x$, $V$ is likely to increase with $x$, and the value of energy $V'$ is likely to decrease when $x$ is high (McNamara and Houston 1986; McNamara 1990; Houston and McNamara 1999). As a result the predation risk that it is optimal to accept in order to obtain a unit of food increases as $x$ decreases. Schuck-Paim et al. note that some experiments that claim to have found irrational behaviour involve treatments that result in differences in state that could explain the results without requiring a violation of rationality.

## 3.3  MODELS OF OPTIMAL BEHAVIOUR IN WHICH TRANSITIVITY IS VIOLATED

In the second sort of explanation, selection acts on behaviour and violations of transitivity result.

### 3.3.1  Options are linked

Navarick and Fantino (1972, 1974, 1975) studied the behaviour of pigeons on the concurrent-chains procedure and concluded that transitivity was violated. The concurrent-chains procedure gives an animal two alternatives, known as initial links. For example, a pigeon might be faced with two discs, one red and one green. The pigeon can peck either disc. There is a chance that a peck on a disc switches the pigeon from the initial link to what is known as a terminal link. This link gives the pigeon some food after the pigeon waits for a given time. In one sort of experiment each terminal link gives a fixed amount of food after a fixed delay. The procedure can be regarded as a version of prey choice in which a terminal link corresponds to a prey type. The proportion $p$ of responses to alternative 1 (i.e. the number of responses made on the initial link for alternative 1, divided by the total number of responses made on both initial links) is taken to be a measure of preference for alternative 1.

Houston et al. (1987) present a model of how responses should be allocated to the initial links if the overall rate of gain is to be maximized. They show that violations of SST can result. The reason for this is that the allocation of responses on the initial links is not a measure of preference for the terminal links. Instead the model is based on the fact that the allocation can influence the probability of which terminal link is obtained, and hence determines the overall reward rate. When faced with terminal links $a$ and $b$, the optimal allocation $p^*$ $(a, b)$ is the one that maximizes the overall rate. In other words, if decisions are based on overall rate, then $p^*$ should be preferred to all other allocations. Note that provided $0 < p^* < 1$, there is a chance of obtaining both terminal links, so both links influence the overall rate. (I have simplified the argument by phrasing it in terms of finding the best allocation. In fact the model finds the best pattern of switching between the initial links. This pattern determines the 'best' allocation.)

Although rate maximization provides a possible explanation for violations of transitivity on concurrent chains, I do not believe that it is the correct one. In many operant experiments, animals do not maximize rate (Mazur 1981; Heyman and Herrnstein 1986). Extensions of descriptive models of choice on concurrent chains can predict that choice does not satisfy SST. For example, Houston (1991) develops a version of the delay-reduction hypothesis (Fantino and Abarca 1985) for terminal links that differ in magnitude of reward and delay. Substitutability does not hold in this model.

### 3.3.2 State-dependent behaviour

Houston et al. (2007b) demonstrate that violations of transitivity can occur when animals make optimal choices that depend on an internal variable such as its energy reserves. For example, an animal might be selected to maximize its probability of surviving the winter. To do so, it must avoid the dangers of starvation and predation. The animal dies of starvation if its energy reserves fall to zero. At each decision time the animal chooses between foraging options that differ in terms of energetic gain and risk of being killed by a predator. A strategy for the animal specifies how the animal's choice of foraging option depends on its energy reserves when the decision is made. Dynamic programming (Houston et al. 1988; Houston and McNamara 1999; Clark and Mangel 2000) is used to find the optimal strategy, i.e. the strategy that maximizes the animal's probability of survival. Houston et al. (2007b) find the optimal strategy in three environments. Each environment consists of two foraging options. The optimal strategy specifies which option should be chosen in each state, under the assumption that the animal will make repeated choices between the options over a long period. Houston et al. then focus on the choices made by the animal when it has a particular level of energy reserves. They show that it can be optimal for the animal to choose option *a* when the environment consists of options *a* and *b*, to choose option *b* when the environment consists of options *b* and *c*, but to choose option *c* when the environment consists of options *a* and *c*. Decisions are perfectly consistent at the level of strategies but appear to violate transitivity if a single choice by an animal with given reserves is viewed as demonstrating that one option is preferred to the other. What this means in the context of experiments that violate transitivity is that the experimenter assumes that the animal is making the best choice between two options, whereas the subject is following a rule that gives it the best performance if it is going to make repeated choices between the options.

### 3.3.3 Errors in decision making

Houston (1997) shows that errors in decision making can result in intransitive choices. The errors have to have a particular pattern in which costly errors are less likely than cheap errors (Kacelnik 1984; Houston 1987). The argument exploits a general feature of decisions between foraging options that differ in terms of energy gained and handling time. An option of type $i$ has energy $e_i$ and handling time $h_i$. From McNamara (1982) and McNamara and Houston (1987), the energetic value of accepting a type $i$ item is

$$H_i = e_i - \gamma h_i ,$$

where $\gamma$ is the rate of energetic gain. It is optimal to accept all item types for which $H > 0$, i.e. for which $e/h > \gamma$ (Houston and McNamara 1999). This shows the importance of the rate of energetic gain $\gamma$ in optimal decision making. What is crucial for optimal choice is not the absolute properties of an option, but its value compared with $\gamma$. (This idea of relative value is clear in Charnov et al. 1981, p. 28.) The argument of Houston is based on the fact that if the decision maker has a choice between two items, because of errors there is a probability that the preferred option will not always be chosen. Following McNamara and Houston (1987), the probability is assumed to be given by

$$p(a,b) = \frac{\exp(\beta H_a)}{\exp(\beta H_a) + \exp(\beta H_b)},$$

where $\beta$ is a positive constant.

As a result, both items have an effect on $\gamma$. In turn $\gamma$ determines the $H$ values, which determine error probabilities. What this means is that the value of an option depends on the overall context of the choice procedure. The same sort of approach can be applied to options that differ in terms of energy and predation.

### 3.3.4 Restricted class of rules

Waksberg et al. (2009) argue that irrational behaviour can maximize fitness. They find by simulation that a rule that does not assign constant values to options can out-perform a rule that uses constant values. I argue that their result is an artefact of not allowing state-dependent decisions.

In the model of Waksberg et al. (2009), a forager makes a series of choices between items. An item is characterized by two attributes $y$ and $z$. An animal's lifetime reproductive success is a function of the total amounts of the two attributes that it has obtained after the last decision has been made. Waksberg et al. (2009) consider three strategies for choosing between options.

(1) An option is chosen at random, i.e. there is no evaluation of the options.
(2) The animal is economically rational (ER): it assigns a value to each option that does not depend on the other available options and the option with the highest value is selected.
(3) The animal adopts a trade-off contrast (TC) rule. Decisions are based on a flexible valuation of the available options that depends in part on these options. This form of rule, which was suggested by Tversky and Simonson (1993), can result in violations of independence of irrelevant alternatives.

Waksberg et al. do not allow decisions to depend on the decision maker's state. This limitation means that they eliminate the optimal strategy. By considering performance when there is just one decision to be made, it can be seen that the optimal choice of option depends on the decision maker's state, as represented by the amount of $y$ and $z$ that it has previously obtained. As a simple example, assume that options are such that $y + z = K$. Let the state of the animal be $(y_0, z_0)$. If it is optimal to maximize $(y_0 + y)(z_0 + z)$, then the best option is given by $y^* = \dfrac{z_0 - y_0 + K}{2}$. Thus in this case, the best option depends on the difference in initial value of the states.

Waksberg et al. show that an animal using a TC rule can sometimes outperform an ER animal. This does not show that it is optimal to be irrational. What it shows is that a flexible strategy that can violate the principles of rational choice is sometimes able to outperform a suboptimal strategy that does not violate these principles. Waksberg et al. justify their decision not to consider state-based strategies, by noting that such strategies can produce violations of transitivity. Although this is true, it does not justify eliminating them when it is necessary to compare various rules with the optimal strategy. I expect that the optimal strategy would produce an evaluation of options that depends on state but, for a given state, meets the requirements of rationality.

## 3.4 DISCUSSION

In this chapter I have confined attention to decisions by single agents, but questions about rational decisions also arise when decisions are made by a group (Franks et al. 2003; Edwards and Pratt 2009; Robinson et al. 2009). I have not attempted to evaluate the extent to which decisions are intransitive. This empirical issue is discussed by Birnbaum and Gutierrez (2007), Birnbaum and Schmidt (2010) and Regenwetter et al. (2011). Instead, my focus has been on the different possible ways in which intransitive, and more generally irrational, behaviour can be given an evolutionary explanation.

A common feature of several of the cases that I have examined is that transitivity appears to be violated because we do not have the correct view of the decision maker's options. I now illustrate this point, with reference to both the concurrent-chains procedure and the case of state-dependent decisions.

The data from the concurrent-chains procedure show violations of transitivity if the relative allocation on the initial links is viewed as a measure of preference for the terminal links. This view may not be correct. As models of the chains procedure show, allocation on the initial links influences the decision maker's expected gain and expected time. In the model of Houston et al. (1987), the decision maker is assumed to be selecting the allocation that results in the highest possible rate of energetic gain. This amounts to viewing the decision maker as choosing between possible allocations. The resulting allocations can appear to violate transitivity if they are regarded as a measure of preference for the terminal links.

The violation of transitivity found by Houston et al. (2007b) is striking because the choices are deterministic rather than stochastic. The decision maker is viewed as selecting the best of all possible state-dependent strategies for choosing between the two available options. For a given pair of options, there is a single state-dependent strategy that maximizes the decision maker's probability of survival. Decisions appear to violate transitivity if we focus on a particular state and view the decision in that state as an indication of preference between the two options. This view distorts the behaviour of the decision maker by taking a single choice out of its proper context. Selection is not acting on single choices between the two available options. Instead it is acting on strategies for choosing between the options. At the level of strategies, there is no violation of transitivity; under particular conditions, there is one best strategy that should be adopted. This strategy specifies which of the two options should be chosen in each possible state. What is crucial is the assumption that the pair of options that are offered in a choice test will continue to be available in the future, i.e. the animal can always choose between the options.

This raises an important general point. In many cases in which an animal is offered a choice, the prediction about the optimal decision depends on whether the choice will be available in the future (cf. McNamara and Houston 1994). The extent to which organisms expect options to persist is likely to have a significant effect on behaviour in many cases. In the model of Houston et al. (2007b), the options indicate future options and the strategy is based on both options being used. Even if it is not going to be selected, by providing information an option can produce behaviour that challenges rationality. An example is provided by cases in which the decision maker does not have perfect knowledge of the two options $a$ and $b$ that are presented. In the absence of information, $a$ is preferred to $b$. When option $c$ is added, it provides information that results in $b$ being preferred to $a$. For accounts that do not depend on information, see Tsetsos et al. (2010). As an aside, the distinction between selection acting on decisions and selection acting on rules for making decisions has been made in the context of game-theoretic interactions. McNamara et al. (1999) show that predictions about the stable level of parental care depend on whether parents adopt a genetically determined level of effort or a rule that determines the level of effort on the basis of their partner's effort.

Tversky (1969) identifies transitivity as 'one of the basic and most compelling principles of rational behavior' (p. 45). He points out that if a decision maker violates transitivity, then the resulting decisions can be used to make the decision maker act as a 'money-pump'. The pump works as follows. Assume that $a > b$ and $b > c$ but $c > a$, where $a > b$ indicates that $a$ is preferred to $b$. If the decision maker has option $c$, then because $b > c$, the decision maker will pay a certain amount to exchange $c$ for $b$. Similarly, because $a > b$, an amount will be paid to exchange $b$ for $a$. Finally, because $c > a$, an amount will be paid to exchange $a$ for $c$, taking the decision maker round a cycle of exchanges to the starting point for a loss of money at each stage of the cycle. For further discussion of noncyclic choices see Gustafsson (2010).

Despite the 'money-pump' argument, Tversky is not convinced that intransitive behaviour is irrational. In an argument that anticipates the 'rule-of-thumb' and 'ecological rationality' approaches to behaviour, Tversky points out that when both the cost of evaluating options and consequences of errors are considered, a simple decision procedure may be best even if it does not necessarily produce transitive decisions. He writes: 'people employ various approximation methods that enable them to process the relevant information in making a decision. The particular approximation scheme depends on the nature of the alternatives as well as on the ways in which they are presented or displayed' (1969, p. 46). Tversky goes on to state that the use of such methods implicitly assumes 'that the world is not designed to take advantage of our approximation methods' (ibid., p. 46). His conclusion is that rules may typically perform well, even though they can produce violations of transitivity in some circumstances.

Violations of transitivity arise because the value of an option depends on the other options that are available, i.e. value is context dependent. In models of choice, this context dependence arises in various ways. In regret theory (Loomes and Sugden 1982), it is based on comparing an outcome with alternatives. In the state-dependent models of Houston et al. (2007b), it arises because both options are taken to be available in the future. If we wish to understand why natural selection produces violations of transitivity, it might be better to focus on the benefits of context-dependent value rather than on the costs of intransitive choices.

It is instructive to compare the case of errors in decision making with the restricted class of rules. In the former case, it is assumed that decision making is subject to error, with costly errors being rare. This assumption seems reasonable. A consequence of the assumption is that both of the two available options influence the rate of gain that the decision maker can expect, and violations of transitivity emerge from the maximization of rate of gain. In contrast, analysis based on the restricted class of rules finds that 'irrational' rules perform well by excluding state-dependent rules. This exclusion does not seem reasonable, and so I do not believe that the analysis provides a convincing argument for natural selection favouring irrational behaviour.

Finally, none of the models that I have considered involve game-theoretic interactions. Incorporating such interactions can result in the evolution of decisions that do not maximize expected gain (Dickson 2008). I believe that further work in this area is likely to be informative.

## REFERENCES

Bateson M. and Healy S. D. (2005). Comparative evaluation and its implications for mate choice. *Trends Ecol. Evol.*, 20, 659–664.

Bednekoff P. A. (2007). Foraging in the face of danger. In: *Foraging* (eds. Stephens D. W., Brown J. S. and Ydenberg R. C.). University of Chicago Press, pp. 305–329.

Birnbaum M. H. (2010). Testing lexicographic semiorders as models of decision making: priority dominance, integration, interaction, and transitivity. *J. Math. Psychol.*, 54, 363–386.

Birnbaum M. H. and Gutierrez R. J. (2007). Testing for intransitivity of preferences predicted by a lexicographic semi-order. *Organ. Behav. Hum. Decis. Process*, 104, 96–112.

Birnbaum M. H. and LaCroix A. R. (2008). Dimension integration: testing models without trade-offs. *Organ. Behav. Hum. Decis. Process*, 105, 122–133.

Birnbaum M. H. and Schmidt U. (2010). Testing transitivity in choice under risk. *Theory Decis.*, 69, 599–614.

Brandstatter E., Gigerenzer G. and Hertwig R. (2006). The priority heuristic: making choices without trade-offs. *Psychol. Rev.*, 113, 409–432.

Browne W. J., Caplen G., Edgar J., Wilson L. R. and Nicol C. J. (2010). Consistency, transitivity and inter-relationships between measures of choice in environmental preference tests with chickens. *Behav. Processes*, 83, 72–78.

Charnov E. L., Los-den Hartogh R. L., Jones W. T. and Van den Assem J. (1981). Sex ratio evolution in a variable environment. *Nature*, 289, 27–33.

Clark C. W. and Mangel M. (2000). *Dynamic state variable models in ecology.* Oxford University Press.

Cooper W. S. (2001). *The evolution of reason.* Cambridge University Press.

Day B. and Loomes G. (2010). Conflicting violations of transitivity and where they may lead us. *Theory Decis.*, 68, 233–242.

Dickson E. S. (2008). Expected utility violations evolve under status-based selection mechanisms. *J. Theor. Biol.*, 254, 650–654.

Edwards S. C. and Pratt S. C. (2009). Rationality in collective decision-making by ant colonies. *Proc. R. Soc. B*, 276, 3655–3661.

Fantino E. and Abarca N. (1985). Choice, optimal foraging, and the delay-reduction hypothesis. *Behav. Brain Sci.*, 8, 315–330.

Fishburn P. C. (1973). Binary choice probabilities: on the varieties of stochastic transitivity. *J. Math. Psychol.*, 10, 327–352.

Franks N. R., Mallon E. B., Bray H. E., Hamilton M. J. and Mischler T. C. (2003). Strategies for choosing between alternatives with different attributes: exemplified by house-hunting ants. *Anim. Behav.*, 65, 215–223.

Gigerenzer G. (2000). *Adaptive thinking.* Oxford University Press.

Green L. and Estle S. J. (2003). Preference reversals with food and water reinforcers in rats. *J. Exp. Anal. Behav.*, 79, 233–242.

Gustafsson J. E. (2010). A money-pump for acyclic intransitive preferences. *Dialectica*, 64, 251–257.

Heyman G. M. and Herrnstein R. J. (1986). More on concurrent interval-ratio schedules – a replication and review. *J. Exp. Anal. Behav.*, 46, 331–351.

Houston A. I. (1987). The control of foraging decisions. In: *Quantitative analyses of behavior VI – foraging* (eds. Commons M. L., Kacelnik A. and Shettleworth S. J.). Lawrence Erlbaum Associates, Hillsdale, NJ, pp. 41–61.

(1991). Violations of stochastic transitivity on concurrent chains – implications for theories of choice. *J. Exp. Anal. Behav.*, 55, 323–335.

(1997). Natural selection and context-dependent values. *Proc. R. Soc. B*, 264, 1539–1541.

(2009). Flying in the face of nature. *Behav. Processes*, 80, 295–305.

Houston A. I., Clark C., McNamara J. and Mangel M. (1988). Dynamic models in behavioural and evolutionary ecology. *Nature*, 332, 29–34.

Houston A. I. and McNamara J. M. (1989). The value of food: effects of open and closed economies. *Anim. Behav.*, 37, 546–562.

(1999). *Models of adaptive behaviour*. Cambridge University Press.

Houston A. I., McNamara J. M. and Steer M. D. (2007a). Do we expect natural selection to produce rational behaviour? *Phil. Trans. R. Soc. B*, 362, 1531–1543.

(2007b). Violations of transitivity under fitness maximization. *Biol. Lett.*, 3, 365–367.

Houston A. I., Sumida B. H. and McNamara J. M. (1987). The maximization of overall reinforcement rate on concurrent chains. *J. Exp. Anal. Behav.*, 48, 133–143.

Hutchinson J. M. C. and Gigerenzer G. (2005). Connecting behavioural biologists and psychologists: clarifying distinctions and suggestions for further work. *Behav. Processes*, 69, 159–163.

Kacelnik A. (1984). Central place foraging in starlings (*Sturnus vulgaris*). 1. Patch residence time. *J. Anim. Ecol.*, 53, 283–299.

(2006). Meanings of rationality. In: *Rational animals?* (eds. Hurley S. and Nudds M.). Oxford University Press, pp. 87–106.

Kalenscher T., Tobler P. N., Huijbers W., Daselaar S. M. and Pennartz C. M. A. (2010). Neural signatures of intransitive preferences. *Front. Hum. Neurosci.*, 4, doi: 10.3389/fnhum.2010.000491967.

Katsikopoulos K. V. and Gigerenzer G. (2008). One-reason decision-making: modeling violations of expected utility theory. *J. Risk Uncertain.*, 37, 35–56.

Latty T. and Beekman M. (2011). Irrational decision-making in an amoeboid organism: transitivity and context-dependent preferences. *Proc. R. Soc. B*, 278, 307–312.

Lima S. L. (1998). Stress and decision making under the risk of predation: recent developments from behavioral, reproductive, and ecological perspectives. In: *Advances in the Study of Behavior: Stress and Behavior*, vol. 27, pp. 215–290.

Lima S. L. and Dill L. M. (1990). Behavioral decisions made under the risk of predation: a review and prospectus. *Can. J. Zool.*, 68, 619–640.

Linares P. (2009). Are inconsistent decisions better? An experiment with pairwise comparisons. *Eur. J. Oper. Res.*, 193, 492–498.

Loomes G. and Sugden R. (1982). Regret theory: an alternative theory of rational choice under uncertainty. *Econ. J.*, 92, 805–824.

Luce R. D. (1959). *Individual choice behavior*. John Wiley, New York.

Luce R. D. and Suppes P. (1965). Preference, utility, and subjective probability. In: *Handbook of mathematical psychology* (eds. Luce R.D., Bush R.R. and Galanter E.). John Wiley, New York, pp. 249–410.

Mazur J. E. (1981). Optimization theory fails to predict performance of pigeons in a 2-response situation. *Science*, 214, 823–825.

McNamara J. M. (1982). Optimal patch use in a stochastic environment. *Theor. Popul. Biol.*, 21, 269–288.

(1990). The policy which maximizes long-term survival of an animal faced with the risks of starvation and predation. *Adv. Appl. Prob.*, 22, 295–308.

McNamara J. M., Gasson C. E. and Houston A. I. (1999). Incorporating rules for responding into evolutionary games. *Nature*, 401, 368–371.

McNamara J. M. and Houston A. I. (1985). Optimal foraging and learning. *J. Theor. Biol.*, 117. 231–249.

(1986). The common currency for behavioral decisions. *Am. Nat.*, 127, 358–378.

(1987). Partial preferences and foraging. *Anim. Behav.*, 35, 1084–1099.

(1992). Risk-sensitive foraging – a review of the theory. *Bull. Math. Biol.*, 54, 355–378.

(1994). The effect of a change in foraging options on intake rate and predation rate. *Am. Nat.*, 144, 978–1000.

(1997). Currencies for foraging based on energetic gain. *Am. Nat.*, 150, 603–617.

(2009). Integrating function and mechanism. *Trends Ecol. Evol.*, 24, 670–675.

Navarick D. J. and Fantino E. (1972). Transitivity as a property of choice. *J. Exp. Anal. Behav.* 18, 3, 389–401

(1974). Stochastic transitivity and unidimensional behavior theories. *Psychol. Rev.*, 81, 426–441.

(1975). Stochastic transitivity and the unidimensional control of choice. *Learn. Motiv.*, 6, 179–201.

Pavlic T. P. and Passino K. M. (2010). When rate maximization is impulsive. *Behav. Ecol. Sociobiol.*, 64, 1255–1265.

Rachlin H. and Green L. (1972). Commitment, choice and self-control. *J. Exp. Anal. Behav.*, 17, 15–22.

Reaney L. T. (2009). Female preference for male phenotypic traits in a fiddler crab: do females use absolute or comparative evaluation? *Anim. Behav.*, 77, 139–143.

Regenwetter M., Dana J. and Davis-Stober C. P. (2011). Transitivity of preferences. *Psychol. Rev.*, 118, 42–56.

Rieskamp J. (2008). The probabilistic nature of preferential choice. *J. Exp. Psychol. Learn. Mem. Cogn.*, 34, 1446–1465.

Robinson E. J. H., Smith F. D., Sullivan K. M. E. and Franks N. R. (2009). Do ants make direct comparisons? *Proc. R. Soc. B.*, 276, 2635–2641.

Royle N. J., Lindstrom J. and Metcalfe N. B. (2008). Context-dependent mate choice in relation to social composition in green swordtails *Xiphophorus helleri. Behav. Ecol.*, 19, 998–1005.

Schuck-Paim C., Pompilio L. and Kacelnik A. (2004). State-dependent decisions cause apparent violations of rationality in animal choice. *PLoS. Biol.*, 2, 2305–2315.

Shafir S. (1994). Intransitivity of preferences in honey-bees – support for comparative-evaluation of foraging options. *Anim. Behav.*, 48, 55–67.

Simon H. A. (1956). Rational choice and the structure of the environment. *Psychol. Rev.*, 63, 129–138.

 (1990). Invariants of human behavior. *Annu. Rev. Psychol.*, 41, 1–19.

Stephens D. W. and Krebs J. R. (1986). *Foraging theory*. Princeton University Press.

Todd P. M. and Gigerenzer G. (2003). Bounding rationality to the world. *J. Econ. Psychol.*, 24, 143–165.

 (2007). Environments that make us smart: ecological rationality. *Curr. Dir. Psychol. Sci.*, 16, 167–171.

Tsetsos K., Usher M. and Chater N. (2010). Preference reversal in multiattribute choice. *Psychol. Rev.*, 117, 1275–1293.

Tversky A. (1969). Intransitivity of preference. *Psychol. Rev.*, 76, 31–48.

Tversky A. and Russo J. E. (1969). Substitutability and similarity in binary choices. *J. Math. Psychol.*, 6, 1–12.

Tversky A. and Simonson I. (1993). Context-dependent preferences. *Manage. Sci.*, 39, 1179–1189.

Waite T. A. (2001). Intransitive preferences in hoarding gray jays (*Perisoreus canadensis*). *Behav. Ecol. Sociobiol.*, 50, 116–121.

Waksberg A., Smith A. and Burd M. (2009). Can irrational behaviour maximise fitness? *Behav. Ecol. Sociobiol.*, 63, 461–471.

# CHAPTER 4

# Evolution, dynamics and rationality
## The limits of ESS methodology

*Simon M. Huttegger and Kevin J. S. Zollman*

### 4.1 INTRODUCTION

Until quite recently, the study of games has developed almost independently in both economics and biology. In both fields, however, focus has been on developing *equilibrium concepts* – stable states that should be expected to emerge as a result of evolution by natural selection acting on populations of organisms (in biological game theory) or from the "rational" choices of individuals (in economics). While one could formalize both the dynamics of natural selection and notions of rationality, the history of game theory has relied more on intuitive characterizations of both, rather than mathematically precise characterizations.

As an example, consider the canonical equilibrium concept, Nash equilibrium. In economics it is argued that if a player expects the other players to play a Nash equilibrium, then this player does best by playing her part of the Nash equilbrium. A Nash equilibrium thus represents a kind of stable state of the behavior and expectations of rational individuals. What comprises these beliefs and how they came about is often left to informal discussion. In biology, a population where all individuals play a Nash equilibrium is similarly stable; since each individual is doing the best she can given what the others are doing, it is assumed that natural selection will not immediately (without perturbation) take the state away from the Nash equilibrium. Again, the process by which natural selection operates, and the process by which a population comes to play a Nash equilibrium is left for informal discussion.

The absence of mathematically precise foundations was not due to ignorance or ambivalence but was instead an attempt at generality. It was

This material is based upon work supported by the National Science Foundation under Grant No. EF 1038456. Any opinions, findings, and conclusions or recommendations expressed in this material are those of the authors and do not necessarily reflect the views of the National Science Foundation.

believed that the intuitive notions on which the fields relied were so general as to be a feature of any formal characterization of either natural
selection or rationality respectively. The result of this intuitive reflection
was believed to be robust across many different potentially inconsistent
precise specifications of the underlying notions. Because of this desire for
robustness and also because of the difficulty of analyzing epistemic models, the study of game theory in economics proceeded by developing more
and more sophisticated equilibrium notions creating a large-scale project
called the "equilibrium refinement" program.

More recently in economics, significant attention has been given to
the foundations of equilibrium concepts used there. The notions of belief,
that had previously been left informal, have been explicitly represented
utilizing a variety of models. This "epistemic program" in economics
focuses on formally explicating the notion of rationality and uncovers
when rational individuals will indeed come to play equilibria (see e.g.
Brandenburger 2007). In addition, many scholars have utilized models of
bounded rationality that attempt to determine the robustness of equilibrium notions when intuitive rationality assumptions are relaxed.[1]

This foundational turn has yet to take hold in biology. In biology, the
predominant equilibrium concept is the notion of an *evolutionarily stable strategy* (ESS) as introduced in Maynard Smith and Price (1973). The
concept of an ESS is also a refinement of Nash equilibrium which better
represents a state which cannot be eliminated by natural selection with
the introduction of small perturbations. As with equilibrium refinements
in economics, ESS provides a short cut to analyzing the specific properties of the dynamics of evolution by natural selection; in fact, historically
it precedes the introduction of dynamical systems to evolutionary game
theory (Taylor and Jonker 1978). ESS allows one to infer the stability
of certain population states without recourse to some specific underlying
dynamics. The claim is that the stability of ESSs will hold across a large
range of evolutionary dynamics. Robustness comes for free.

ESS, therefore, is a powerful tool to analyze evolutionary phenomena.
Though it may be tricky to find ESSs in particular cases, it is generally
much easier than studying an underlying dynamical system. Furthermore,
ESS goes beyond the claim of stability for a particular dynamical system.
That is, ESS allows one to make a more general claim than mere dynamic

---

[1] It should be noted that John Nash justified his concept using this notion of learning among
boundedly rational agents. However, this justification was largely lost until this methodology
became popular much later.

|   | C | D |
|---|---|---|
| C | 2<br>2 | 3<br>0 |
| D | 0<br>3 | 1<br>1 |

Figure 4.1. A prisoner's dilemma.

stability. Thus, methods based on ESS appear to be more tractable and seem to deliver more general results than is possible by investigating stability in evolutionary dynamical systems. For these reasons, ESS is a much used concept in the biological literature on evolutionary games.

In this chapter, we shall argue that in many relevant cases there are non-ESS states that are nevertheless evolutionarily significant (Section 4.4). This implies that attempting to find ESSs will often not be sufficient to obtain an adequate understanding of an evolutionary process. In order to make this case, we shall distinguish between ESS as a concept and ESS as a methodology (Section 4.2). By the latter we roughly mean the maxim that finding the ESSs of the game allows one to derive all, or nearly all, relevant conclusions about the game in question.

The second topic of this chapter is to relate the ESS concept to certain equilibrium concepts of economic game theory (Section 4.3). Like ESS, these equilibrium concepts are refinements of Nash equilibria. Our distinction between the concept itself and the methodology based on it (that its implications are the only relevant ones) carries over to the case of the other refinements. We conclude that, both in the case of ESS and in the case of other Nash equilibrium refinements, the methodology fails in important cases.

## 4.2 ESS AND ESS METHODOLOGY

We start by briefly reviewing the ESS concept (Maynard Smith and Price 1973; Maynard Smith 1982). Suppose a symmetric finite two-player game $\Gamma$ is being played repeatedly in a panmictic population; i.e. the population is effectively infinite, and interactions are happening at random (there is no population structure) and are taking place between two individuals whose roles in the game cannot be distinguished. Let $u(s,s')$ denote the payoff strategy $s$ obtains when interacting with strategy $s'$ (payoff may be understood as incremental fitness). If $s^*$ is a strategy of $\Gamma$, then $s^*$ is evolutionarily stable if for all strategies $s$ of $\Gamma$ other than $s^*$, the following two conditions hold:

(1) $u(s^*,s^*) \geq u(s,s^*)$

(2) If $u(s^*,s^*) = u(s,s^*)$, then $u(s^*,s) > u(s,s)$.

The first of these conditions states that $s^*$ has to be a symmetric Nash equilibrium. The second condition requires $s^*$ to obtain a higher payoff when interacting with $s$ than $s$ gets when interacting with itself whenever $s$ is an alternative best response to $s^*$.

As an illustration consider the prisoner's dilemma, pictured in Figure 4.1. In this game, strategy $D$ is an ESS, because it is a Nash equilibrium (satisfying the first condition) and because all other strategies do worse against $D$ than $D$ does against itself (rendering the second condition void). $C$ is not an ESS, because it is not a Nash equilibrium. There are many other applications of ESS found in the biological literature, beginning with the Hawk–Dove game (Maynard Smith and Price, 1973). Finding ESS in larger games can often be a nontrivial matter.

In an evolutionary setting, ESS can be understood the following way. Suppose the incumbent strategy $s^*$ is being played by the whole population. If a small number of mutants playing $s$ arise and condition (1) does not obtain, then clearly the population state $s^*$ cannot be stable, for $s$ will take over the population when it is given enough time. If, on the other hand, (1) obtains with a strict inequality sign, then $s$ will not be able to displace the incumbent $s^*$. Condition (2) takes care of the third possibility, where $u(s^*,s^*) = u(s,s^*)$. Here, interactions with the incumbents have no effect on the population. Thus, for $s^*$ to be stable in this case, it needs to get a higher payoff against the mutant than the mutant gets when interacting with itself.

The ESS concept thus readily lends itself to an evolutionary interpretation. Formally, ESS is a refinement of Nash equilibrium. Condition (1) requires an ESS to be a Nash equilibrium, and condition (2) is not fulfilled by all Nash equilibria. ESSs are attractive candidates for evolutionary outcomes because they are dynamically stable in several important evolutionary dynamics. For instance, in the replicator dynamics, a strategy $i$'s per capita growth rate in a panmictic population is given by the difference of its fitness and the average fitness in the population:

$$\dot{x}_i = x_i(f_i - \bar{f}) \tag{4.1}$$

($f_i$ and $\bar{f}$ depend linearly on the payoffs of the game $\Gamma$; see Hofbauer and Sigmund [1998] for details.) This implies that strategies with above-average fitness increase in frequency while strategies with below-average fitness decrease in frequency. It is well known that an ESS is an asymptotically stable state of the replicator dynamics. This means that populations where a sufficiently high frequency of individuals already play the ESS will tend toward a state where all individuals will do so. It should be emphasized that this holds, not only for the replicator dynamics, but for larger classes of dynamics that contain the replicator dynamics. Let us just mention monotone selection dynamics (Hofbauer and Sigmund 1998). Monotone selection dynamics are given by a system of equations

$$\dot{x}_i = x_i g_i(x),$$

where $x$ is the state of the population. For this system to be a monotone selection dynamics it is required that $g_i(x) > g_j(x)$ if, and only if, $f_i > f_j$; i.e. $i$'s growth rate is higher than $j$'s growth rate if, and only if, $i$'s fitness is higher than $j$'s fitness. It is well known that asymptotic stability of ESSs will also hold for monotone selection dynamics (Weibull 1995).

There is one important aspect of ESSs that we have not mentioned so far. The ESS concept does not just apply to the pure strategies of a game as presented above. It can be straightforwardly extended to mixed strategies. There is a certain ambiguity concerning mixed strategy ESSs that won't concern us here (the ambiguity has to do with whether one identifies a mixed strategy ESS as a population state or as a strategy being played by the population).

For our purposes it is more important that in many kinds of interactions ESSs can be identified with the strict Nash equilibria of a game. To be more specific, whenever we start with a two-player asymmetric game (i.e. a game where the two player roles are distinguishable) we may consider the *symmetrized* game based on it (Cressman 2003). In the symmetrized game each player is assumed to be in one of the two player positions of the asymmetric game with equal probability. By taking the payoffs to be expected values, the resulting game is a symmetric game to which the ESS concept can be applied. It turns out that there is a simple relationship between the strict Nash equilibria of the asymmetric game and the ESSs of the corresponding symmetrized game: a strategy of the symmetrized game is an ESS if, and only if, the corresponding pair of strategies is a strict Nash equilibrium of the asymmetric game (Selten 1980). Notice that this result is important for many biological examples. It is relevant whenever the situation at hand allows one to clearly distinguish between two player roles, such as male and female, owner and intruder, parent and offspring, and so on. We follow other authors and use ESS and strict Nash equilibrium interchangeably when talking about asymmetric games. (The two do have to be distinguished when talking about symmetric games. In what follows, no ambiguity should arise, however.)

The ESS concept is a purely formal concept that can be applied whenever one has a finite symmetric two-player game. We think it is useful to distinguish what we call *ESS methodology* from the ESS concept proper. ESS methodology has to do with how to interpret the ESS concept rather than its mathematical definition. ESS methodology consists in (i) describing an evolutionary phenomenon in terms of a game, (ii) finding the ESS or the ESSs of the game and (iii) identifying the ESS or the ESSs of the game with the possible evolutionary outcomes of the situation in question.

While few have explicitly recommended the ESS methodology, it appears to be widespread and implicit in many applications of game theory in evolutionary biology. Maynard Smith cautions against the ESS methodology as a fully general methodology, but uses it himself in some particular cases that we believe are problematic (see Section 4.4). Although they never explicitly recommend the ESS methodology, Searcy and Nowicki (2005, ch. 2) refer only to ESS results when surveying the literature on signaling between relatives. This is not to criticize Searcy and Nowicki, for there is very little else in the literature other than ESS results. Instead we wish to illustrate that, whether or not it is ever explicitly defended, there is an implicit reliance in the biological literature on the ESS methodology.

Notice that some results that we have mentioned above lend some credibility to the ESS methodology. In particular, the stability of ESSs (or, for that matter, strict Nash equilibria in asymmetric games) for several standard evolutionary dynamics certainly seems to be a strong reason to trust step (iii). In this chapter, we show that in certain situations this conclusion is not warranted. The line of argument we will use is to establish the evolutionary significance of non-ESS states (already for standard evolutionary dynamics); we also argue that these non-ESS states are not just odd cases but should be expected to be widespread. There are other limits of the ESS methodology that will be worked out in more detail elsewhere (Huttegger and Zollman, forthcoming).

## 4.3  EQUILIBRIUM REFINEMENTS

For what follows it will be important to point out a similarity between ESS and virtually all equilibrium refinements from economics. This similarity

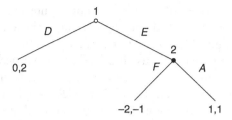

Figure 4.2. The chain-store game.

holds despite the fact that the motivation behind equilibrium refinements is quite different from the motivation behind ESS. As pointed out in the introduction, the general methodology is quite similar. One begins with an intuitive notion – of evolutionary stability in the case of biology and of rationality in the case of economics – and attempts to represent that intuitive notion formally. But the similarity is stronger than this. As we will see, both equilibrium refinements and ESS rely on some notion of perturbation from an equilibrium.

The goal of the equilibrium refinement research project is to rule out certain Nash equilibria as being unreasonable, in the sense that the corresponding Nash equilibrium profile would not be chosen by rational players. Consider *subgame perfect* Nash equilibria as a paradigmatic example of an equilibrium refinement. What is meant by a subgame perfect Nash equilibrium is best explained by example. In the chain-store game (Selten 1978; see Figure 4.2) there are two players. The first player can enter a market that is dominated by a chain store (the second player). If Player 1 enters the market, then Player 2 has to decide between fighting Player 1 or sharing the market. In this case, it is assumed that Player 2 gets the higher payoff from not fighting (as does Player 1). Player 1 can also choose not to enter the market in the first place. In this case, nothing changes. This is clearly the most-preferred outcome from the point of view of Player 2. Player 1 is assumed to have a higher payoff in this outcome than in the outcome where Player 1 chooses to fight. But it would be best for Player 1 if she entered the market and Player 1 did not fight.

Why might Player 1 decide not to enter the market? Well, Player 2 might threaten Player 1 that she will fight, should Player 1 enter the market. Is this threat credible? Selten (1965, 1975, 1978) thinks not. If Player 2 had to decide between fighting and not fighting, she would choose not to fight since this outcome carries the higher payoff.

This kind of reasoning leads to the concept of a subgame perfect Nash equilibrium. There are two pure strategy Nash equilibria in the chain-store game. The first one calls for Player 1 not to enter the market and for Player 2 to fight. In the second one Player 1 enters the market and Player 2 chooses not to fight. Only the latter Nash equilibrium is subgame perfect. It is subgame perfect because it calls for rational play in every subgame of the chain-store game. There are only two subgames, the whole game and the subgame where Player 2 decides to fight or not to fight. "Enter" (Player 1) and "not fight" (Player 2) is rational in both subgames. The first Nash equilibrium is not rational in the second subgame, since it requires Player 2 to fight if she had to decide.

Thus, subgame perfectness rules out the first Nash equilibrium as not being stable. The logic applied here requires us to imagine the two players are choosing according to some Nash equilibrium. This Nash equilibrium will be subgame perfect if all players act rationally when called upon to act even at unreached decision nodes (information sets). We may view this as a perturbation of the players' actions from equilibrium, similar to the population being slightly perturbed from a state in determining whether it is an ESS. In fact, Selten has advanced a theory of ESSs in extensive-form games (such as the chain-store game) along these lines (Selten 1983).

In the case of subgame perfect Nash equilibria, the perturbation from equilibrium is counterfactual. The players are to imagine what would happen if they were called upon to decide at a decision node that is not reached by equilibrium play. This can be turned into a real perturbation, leading to an equilibrium refinement that is closely related to subgame perfect equilibria. It is called a *trembling-hand* perfect equilibrium (Selten 1975). Consider again the chain-store game. Suppose Player 1 decides not to enter the market, but that she cannot implement her strategy with absolute certainty. That is to say, she sometimes makes a mistake and chooses to enter the market (the arbitrary nature of the mistake being captured by the metaphor of a trembling hand). In this situation, it is not rational for Player 2 to choose to fight with high probability. She should choose fighting with a sufficiently low probability. Given that Player 2 is choosing to fight with very low probability, Player 1 does better by entering the market. Thus, taking the probabilities of mistakes to zero, there is no sequence of perturbed Nash equilibria converging to the Nash equilibrium where Player 1 does not enter and Player 2 fights.

The logic of trembling-hand perfect equilibria thus goes as follows. A player assumes that a certain Nash equilibrium is played. Instead of the original game, she imagines that a perturbed game is played where players can only choose mixed strategies. The original Nash equilibrium is trembling-hand perfect if there is a sequence of Nash equilibria of perturbed games converging to the original Nash equilibrium as the probabilities of mistakes go to zero. The subgame perfect equilibrium of the chain-store game passes this test, while the other Nash equilibrium does not. For our purposes, it is important to notice that, again, the players are assumed to be at an equilibrium, and the stability of the equilibrium is determined by considering perturbations from equilibrium (in this case, players making mistakes).

Virtually all equilibrium refinements assume that the players choose according to some particular Nash equilibrium and investigate whether the resulting strategy profile is feasible in terms of some particular perturbation (van Damme 1991). Notice that this has a very similar structure to ESS. Firstly, ESS itself is a Nash equilibrium refinement in the sense that an ESS is also a Nash equilibrium, while not any Nash equilibrium is also an ESS. Moreover, the standard interpretation of ESS also assumes that players are at equilibrium, where now the players' choices are identified with the state of a population. A Nash equilibrium is an ESS if it is stable under a particular perturbation, to wit, the introduction of a low share of a mutant strategy. Thus, ESSs can be viewed as having the same structure as many other equilibrium refinements. The essential aspect appears to be a perturbation away from an equilibrium state.

## 4.4 NON-ESS STATES

As we have just mentioned, from the dynamic stability of ESSs one may be led to the conclusion that one should identify the set of significant evolutionary outcomes of a game with the game's ESSs. One way to undermine this position is by establishing that there are strategies that are not evolutionarily stable but are, nonetheless, stable for the replicator dynamics. Moreover, it is important that this does not only hold for exceptional games (i.e. games with very special payoff parameters), and that these games should be of biological importance. As a first step, we will present just such a game.

The example we have chosen is the well-known Sir Philip Sidney (SPS) game (Maynard Smith 1991). This game was introduced by John Maynard Smith as a simple illustration of the handicap principle (Zahavi 1975; Grafen 1990). The SPS game is a signaling game that allows for conflicts of interest between the sender and the receiver. The sender can be in two states, needy or healthy. Given any state, she can send a message *m* or abstain from doing so. The receiver can transfer a resource to

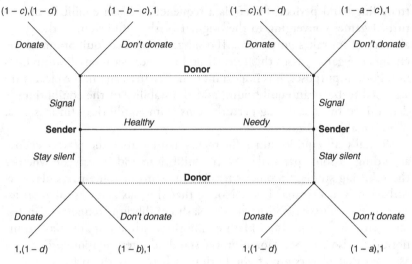

Figure 4.3. The Sir Philip Sidney game. $1 - a$ and $1 - b$ represent the individual fitnesses of the needy and healthy types (respectively) if they do not receive the resource, $c$ is the fitness cost to the sender for sending the signal. $1 - d$ represents the individual fitness of the receiver if she transfers the resource to the sender.

the sender or she can keep it to herself (after observing whether the sender has decided to transmit *m*). An extensive-form representation of the SPS game is pictured in Figure 4.3.

In the version pictured here the receiver's dominant strategy is to keep the resource to herself, because she gains nothing from the transfer. To create the possibility for transfer we can assume that the sender and the receiver can be related to some degree $k \in [0,1]$; $k$ determines the inclusive fitness of a player, i.e. a player's payoff in an interaction is the payoff from her own role, plus $k$ times the payoff from the role of the other player. If $k$ is sufficiently close to 0, the receiver will never donate the resource to the sender, since doing so is detrimental to her own payoff. However, if $k$ is sufficiently large, then it becomes possible for the resource to be transferred in equilibrium.

One population state, the so-called *signaling ESS*, has attracted particular attention. Since the SPS game is an asymmetric game, the signaling ESS really is a strict Nash equilibrium. In the signaling ESS, the sender sends *m* if she is needy and does not send *m* otherwise. The receiver transfers the resource if she gets the signal and keeps it to herself otherwise. The signaling ESS is important because it illustrates the handicap principle, which asserts that signals need to be costly in order for signal reliability to be maintained when there is conflict of interest between a sender and a receiver. The conditions under which the signaling ESS exists reveal the handicap principle. If $d$ is the amount a receiver can transfer, $a$ and $b$ are the gains from a transfer for a needy and healthy sender, respectively, and $k$ is the degree of relatedness between the sender and the receiver, then the signaling ESS exists only if

$$a \geq c + kd \geq b.$$

The second inequality guarantees that a healthy sender will not send the signal in order to obtain the resource. The first inequality guarantees that the gain is sufficiently high for a needy signaler to pay the cost for sending the signal. (There is another condition for the signaling ESS that pertains to the receiver; see Bergstrom and Lachmann [1997] and Huttegger and Zollman [2010] for details.)

So far so good. But the SPS game has other Nash equilibria which do not correspond to ESSs, i.e. which are not strict Nash equilibria. Firstly, there are so-called *pooling equilibria*. These equilibria are characterized by there being no information transfer between the sender and the receiver. There are two kinds of pooling equilibria. In both kinds of pooling equilibria the sender never signals. In one kind of pooling equilibrium the receiver always transfers, while in the other kind she never transfers the resource. One kind of pooling equilibrium always exists (regardless of the parameter settings). But generically it is not the case that both kinds of pooling equilibria exist at the same time.

In Huttegger and Zollman (2010) we show that pooling equilibria cannot be strict, because they are part of Nash equilibrium components. This is quite easy to see. In both kinds of pooling equilibria the sender never sends the signal regardless of whether she is needy or not. Therefore, the receiver's strategies of never transferring the resource or only transferring the resource if the signal is sent are behaviorally equivalent. Similarly, the receiver's strategies of always transferring the resource and only transferring the resource if she does not receive the signal are behaviorally equivalent in this case. This implies that each kind of pooling equilibrium is actually part of a convex set of Nash equilibria and cannot, therefore, be strict.

Pooling equilibria provide a first case of non-ESS states in the SPS game. A second example is given by a so-called *hybrid* or *polymorphic equilibrium*. In a polymorphic equilibrium the sender mixes between two states, "always signal" and "signal only when needy." This means that the sender sometimes reliably indicates her true state and sometimes she does not. The receiver also mixes between two strategies, "never transfer" and "transfer only upon receiving the signal." As a reaction to the sender's dishonesty, the receiver must place some positive weight on not transferring the resource under any circumstances. But, since the sender sometimes signals truthfully, she also has to place some positive probability weight on transferring the resource as a reaction to receiving the signal.

Using the notation introduced above, a polymorphic equilibrium exists only if

$$b - kd > c$$

(Huttegger and Zollman 2010). This shows that the signaling ESS and the polymorphic equilibrium cannot exist together. In fact, the polymorphic equilibrium appears once the signaling ESS ceases to exist, because the signal cost $c$ gets too small. More specifically, it can be shown that $c^* = b - kd$ is the minimum signal cost that guarantees the reliability of the signal (Bergstrom and Lachmann 1997). It follows that in the polymorphic equilibrium there is some information transfer with possibly low costs, despite the fact that there is conflict of interest between the sender and the receiver.

It is clear that the polymorphic equilibrium cannot be an ESS. Since it is not a pure strategy equilibrium (both the sender and the receiver place positive probability weights on two strategies) it cannot be a strict equilibrium. So among the three kinds of equilibria that have been found in the SPS game, only one is an ESS. At this point the question arises whether following the ESS methodology is good advice, i.e. whether we should discard pooling equilibria and the hybrid equilibrium as evolutionarily insignificant.

In Huttegger and Zollman (2010) we show, firstly, that both pooling equilibria and hybrid equilibria are Liapunov stable (when they exist). Liapunov stability is a weaker stability condition than asymptotic stability. It requires solution trajectories that start nearby the state to stay close for all future times. Solution trajectories need not converge (as they do in the case of an asymptotically stable state). Pooling equilibria are not asymptotically stable, because they are elements of a set of equilibria. It can be shown analytically that almost all nearby population states converge to some pooling equilibrium. The hybrid equilibrium is Liapunov stable in a quite different sense. One can show that it is a spiraling center. This means that almost all nearby solution trajectories approach the face on which the hybrid equilibrium can be found while also cycling toward it.

We also note that the signaling ESS does not have a large basin of attraction whenever there is significant conflict of interest between sender and receiver. Most initial populations converge to a state that corresponds to a partial pooling equilibrium. There is a second noteworthy result in Huttegger and Zollman (2010). The hybrid equilibrium, when it exists, appears to have a significantly larger basin of attraction than the signaling ESS in similar cases of conflicting interests. It follows that the ESS methodology fails to a large extent in the SPS game. The signaling ESS, which is the only ESS of this game, fails to have a sufficiently large basin of attraction in the replicator dynamics. It therefore does not appear to be a good prediction of what the evolutionary outcome is in something similar to the SPS game. The two other kinds of Nash equilibria known to exist in the SPS game, the hybrid equilibrium and pooling equilibria, seem to be evolutionarily more significant than the signaling ESS despite their not being ESSs.

These conclusions of course rest on using the replicator dynamics as a model for evolution. It is not clear at this stage what happens when one considers significantly different kinds of dynamics (finite populations, structured populations). To our knowledge there are no general results concerning the stability of ESSs in these models. Therefore, the ESS methodology is best suited when one has a model of evolution similar to the replicator dynamics in mind.

If we stay with the replicator dynamics, can we say a bit more about methodology other than that ESS methodology fails in the SPS game? There are two aspects of this question. Firstly, we chose the SPS game as a biologically relevant example of a game where an ESS methodology fails. Do we expect there to be other examples, or is the SPS game an odd exception? Secondly, is there an equilibrium concept other than ESS that leads to a more suitable methodology? We try to answer these questions in turn in the next section.

### 4.5 EXTENSIVE-FORM GAMES

As to the first question we would like to emphasize that we expect the failure of the ESS methodology to be quite common. The SPS game is an example of a game with a nontrivial extensive-form structure. By this we mean that it is not a simultaneous-move game where all players have to make a choice without any (partial) information about aspects of the choices of the other players. In the SPS game, the receiver knows the action choice of the sender (she receives the sender's signal). In the chain-store game, Player 1 knows if Player 1 has entered the market. Thus, in games with a nontrivial extensive-form structure the sequence of decisions plays an essential role in determining the outcome of the game.

As has been shown elsewhere, games with a nontrivial extensive-form structure often lead to the kinds of equilibria observed in the SPS game (see Cressman 2003; see also Huttegger 2009, 2010). To be more specific, the kind of behavioral equivalences mentioned above, such as "transfer" and "transfer only if signal" given the receiver strategy "don't signal," will be very common in such games. This is simply due to the fact that the sequence of decisions is important in games with a nontrivial extensive form. In many cases this will lead to several of a player's strategies being behaviorally equivalent, given earlier choices of other players. This is important because it influences the topological structure of Nash equilibria. As in the case of pooling equilibria in the SPS game, behaviorally equivalent strategies create a Nash equilibrium component where there are other Nash equilibria arbitrarily close to any particular Nash equilibrium. But such a Nash equilibrium cannot correspond to an ESS, since it is not a strict Nash equilibrium (by the existence of a behaviorally equivalent strategy).

As another example, consider the Nash equilibrium in the chain-store game where Player 1 does not enter and Player 2 chooses to fight (which is neither subgame perfect nor trembling-hand perfect). Given Player 1's choice, both choices of Player 2 are behaviorally equivalent. This implies that the Nash equilibrium is one element of a Nash equilibrium component. Each mixed strategy profile in this component is a Nash equilibrium as long as Player 2 chooses to fight with sufficiently high probability.

Although these kinds of Nash equilibria do not correspond to ESSs, they do correspond to *neutrally stable strategies* (NSSs). For an NSS, the second ESS condition is weakened to a nonstrict inequality. This essentially means that there is no mutant strategy that can take over the population, but that there may be a mutant strategy that can exist alongside the strategy in question. A pooling equilibrium and the non-subgame perfect equilibrium of the chain-store game provide examples of NSSs. (There is again a qualification here. NSS, like ESS, only applies to symmetric games. When we talk about the NSSs of an asymmetric game like the SPS game, we also mean the corresponding NSS of the symmetrized game. There is a precise correspondence between the two, just as in the case of ESS [Cressman 2003].)

In the replicator dynamics, NSS implies Liapunov stability, i.e. population frequencies starting close to an NSS will stay nearby (but don't necessarily converge to the NSS). Thus, one reason why the ESS methodology fails is that it does not take into account NSSs. Should we therefore replace ESS methodology by an NSS methodology? We think not. Taking NSSs into account is a step in the right direction, but one may still fail to capture the stability properties of evolutionary dynamics. NSS implies Liapunov stability, but the converse can fail (Bomze and Weibull 1995). Thus there can be Liapunov stable states in the replicator dynamics that are not NSSs. Furthermore, there are many NSSs which are, in fact, evolutionarily insignificant, as illustrated by the degenerate game where all strategies do equally well. For such a game, any evolutionary dynamics would just drift around and no state could be called stable. We shall not explore the significance of these results, but they should serve to caution us against positing NSS methodology in a strict sense.

The prevalence of Nash equilibria similar to the hybrid equilibrium is much harder to judge. The hybrid equilibrium is a so-called *Nash–Pareto pair* (Hofbauer and Sigmund 1998). This essentially means that it is a mixed equilibrium that is very similar to a strict Nash equilibrium. But there are not many examples of such Nash equilibria in the literature. We suspect that one reason for this has to do with the ESS methodology, which makes it difficult to discover these other kinds of equilibria. We conjecture that it plays an important role in interactions that have some zero-sum aspect to it (just as in the case of the SPS game).

## 4.6 ESS, EQUILIBRIUM REFINEMENTS AND A DYNAMIC PERSPECTIVE

As we have argued in the previous two sections, we think that the ESS methodology fails because it does not properly take into account the dynamic possibilities one already has in baseline dynamical models like the replicator dynamics (which is really tailored towards the ESS concept). We think that there is a deeper reason for this. As explained above, ESS shares with other Nash equilibrium refinements that one starts with what happens to an equilibrium when it is subject to some kind of small perturbation (mutants, trembles, and so on). This can at best capture part of the dynamics under consideration, be it evolutionary dynamics, learning dynamics or the deliberational dynamics of a rational player, namely, the dynamics close to equilibrium. It does not answer the question how players get to an equilibrium if they are far from any equilibrium. (*Far away* may refer to population proportions or to beliefs of players, for example.) In certain special cases this question might not be relevant. For instance, if you have an interior ESS in the replicator dynamics, all interior population states will converge to it. But in many cases this question will be relevant. There could be more than one ESS. Or there could be additional Liapunov stable states that complicate the overall dynamical picture.

One way to address these questions is by determining the basins of attractions for each different equilibrium state. It should be noted that being an ESS does not guarantee a large basin of attraction or even a larger basin of attraction than non-ESS states. For instance, in Huttegger and Zollman (2010), we show that pooling equilibria have a larger basin of attraction than the signaling ESS (if there is conflict of interest between the sender and the receiver), by performing simulations.

While the initial motivations behind the ESS methodology (like the equilibrium refinements program in economics) were noble – theoretical simplicity combined with generality – we believe that this methodology is ultimately inappropriate. The ESS methodology is without a doubt theoretically more simple than an explicitly dynamic methodology, but one must pay too high a cost to secure this simplicity. Many non-ESS states are evolutionarily significant – sometimes more significant than ESS states. Instead we believe that in many cases one must turn to a dynamic methodology for analyzing strategic interactions in biology.

## REFERENCES

Bergstrom, C. T. and Lachmann, M. 1997. "Signalling among relatives I: Is costly signalling too costly?" *Philosophical Transactions of the Royal Society B*, 352: 609–617.

Bomze, I. and Weibull, J. 1995. "Does neutral stability imply Lyapunov stability?" *Games and Economic Behavior*, 11: 173–192.

Brandenburger, A. 2007. "The power of paradox: Some recent developments in interactive epistemology," *International Journal of Game Theory* 25: 465–492.

Cressman, R. 2003. *Evolutionary Dynamics and Extensive Form Games*. MIT Press, Cambridge, MA.

Grafen, A. 1990. "Biological signals as handicaps," *Journal of Theoretical Biology*, 144: 517–546.

Hofbauer, J. and Sigmund, K. 1998. *Evolutionary Games and Population Dynamics*. Cambridge University Press.

Huttegger, S. M. 2009. "On the relation between games in extensive form and games in strategic form," in A. Hiecke and H. Leitgeb (eds.) *Reduction and Elimination in Philosophy and the Sciences*, pp. 375–385. Ontos Press, Frankfurt.

   2010. "Generic properties of evolutionary games and adaptationism," *Journal of Philosophy*, 107: 80–102.

Huttegger, S. M. and Zollman, K. J. S. 2010. "Dynamic stability and basins of attraction in the Sir Philip Sidney game," *Proceedings of the Royal Society B*, 277: 1915–1922.

   (forthcoming) "Methodology in biological game theory," *British Journal for the Philosophy of Science*.

Maynard Smith, J. 1982. *Evolution and the Theory of Games*. Cambridge University Press.

   1991. "Honest signalling: the Philip Sidney game," *Animal Behavior*, 42: 1034–1035.

Maynard Smith, J. and Price, G. 1973. "The logic of animal conflict," *Nature*, 146: 15–18.

Moran, P. A. P. 1962. *The Statistical Processes of Evolutionary Theory*. Clarendon, Oxford.

Searcy, W. and Nowicki, S. 2005. *The Evolution of Animal Communication*. Princeton University Press.

Selten, R. 1965. "Spieltheoretische Behandlung eines Oligopolmodells mit Nachfrägetragheit," *Zeitschrift für die Gesamte Staatswissenschaft*, 121: 301–324, 667–689.

   1975. "Re-examination of the perfectness concept for equilibrium points in extensive games," *International Journal of Game Theory*, 4: 25–55.

   1978 "The chain store paradox," *Theory and Decision*, 9: 127–159.

   1980. "A note on evolutionarily stable strategies in asymmetrical animal conflicts," *Journal of Theoretical Biology*, 84: 93–101.

   1983. "Evolutionary stability in extensive two-person games," *Mathematical Social Sciences*, 5: 269–363.

Taylor, P. D. and Jonker, L. 1978. "Evolutionarily stable strategies and game dynamics," *Mathematical Biosciences*, 40: 145–156.

van Damme, E. 1991. *Stability and Perfection of Nash Equilibria*. Springer, New York.

Weibull, J. 1995. *Evolutionary Game Theory*. MIT Press, Cambridge, MA.

Zahavi, A. 1975. "Mate selection – The selection of a handicap," *Journal of Theoretical Biology*, 53: 205–214.

CHAPTER 5

# *Are rational actor models "rational" outside small worlds?*

*Henry Brighton and Gerd Gigerenzer*

## 5.1 INTRODUCTION

Given a formally well-defined task, a rational actor model defines a rationally justified, optimal response. Rational actor models are desirable goals in the behavioral, cognitive, and social sciences, but in this chapter we use the distinction between small and large worlds to question the cachet associated with the terms "rational" and "optimal." Ideally suited to the analysis of small world problems, both concepts can be counter-productive in the analysis of large world problems. In small worlds, the relevant problem characteristics are certain and uncontroversial in their formalization. For example, a tin can manufacturer seeking to minimize the tin used to package 12 ounces of soup might use solid geometry to determine an optimal can design. In this small world the manufacturer is safe in claiming that no other can design uses less tin. Large worlds are characterized by inherent uncertainty and ignorance, properties which undermine the validity and existence of optimal responses. An aircraft manufacturer designing a flight control system, for instance, faces a large world problem due to the complexity and uncertainty of the operating conditions. Rational actor models may rest on rigorous formal foundations, but they can also signify the questionable use of small world methods to understand large world problems.

With similar concerns, Savage introduced the distinction between small and grand worlds when assessing the limits of Bayesian decision theory (Savage 1954). Savage's decision theory for small worlds – those where an agent has access to a decision matrix defining states of the world, consequences, and actions – shows that the agent will maximize subjective expected utility, providing their preferences satisfy Savage's axioms. Savage saw the problem of casting large world problems, those involving uncertainty and ignorance, in these terms as "utterly ridiculous" (p. 16). Consequently, we will use the term "Savage's problem" to refer to obstacles and potential dangers in using analytic methods geared for small worlds to theorize, and make statements about, large worlds. Specifically, we consider Savage's problem in the context of inductive inference, where a decision maker is required to generalize from observations and infer statistical properties of the environment. For example, an organism foraging for food may infer regularities in the distribution of food items from observations of previous food items. What distinguishes small from large worlds in inductive inference? Moreover, how significant is Savage's problem to the study of inductive inference, where rational actors and optimal responses play a particularly influential role?

We will examine these questions by first setting out the relationship between inductive inference, uncertainty, and error, using statistical learning theory. Within this setting we consider barriers to the identification of optimal responses. The ubiquity of these barriers in the study of real-world problems places a question mark over the range of applicability of rational actor models, and highlights the need for alternative approaches to studying how they function. We then use the study of simple heuristics to illustrate an approach based on algorithmic modeling and competitive model testing, reflecting a relatively recent movement in statistics and pattern recognition (Breiman 2001). Well-informed theorists know that good models hinge on insightful abstraction, and the assumptions made in order to formalize a model will be breached. These are basic facts about models in general. Using the distinction between small and large worlds, our goal is to understand the limits of a specific class of model, rational actor models, which often discharge uncertainty in order to arrive at an optimal solution.

## 5.2 UNCERTAINTY IN INDUCTIVE INFERENCE

All organisms face uncertainty. For instance, look closely at any aspect of cognition and it will likely involve the process of inductive inference, the problem of identifying systematic patterns in observations. Everyday examples include inferring the properties of a visual scene, inferring the intentions of a speaker from a single utterance, or deciding if it is quicker to take the bus, bicycle, or train. Because an infinite number of explanations will always be compatible with a finite series of observations, each of these problems involves uncertainty. Two key questions guide the study of inductive inference in humans and other animals. First, how, in mechanistic terms, do organisms arrive at inductive inferences? Second, what is the relationship between the behavior we observe in organisms, and the optimal behavior, as defined by a rational actor model? We will begin by addressing the second question, and start by taking an idealized perspective on uncertainty, where we adopt the the perspective of an omniscient observer with full knowledge of the task at hand. Relative to this omniscient observer, our first task is to categorize the various forms of uncertainty agents face when making inductive inferences. This will require making the inductive inference problem more precise.

### 5.2.1 Learning from examples

Consider an agent interacting with a series of partners in some game-theoretic setting. Prior to interacting with a new partner, it would be useful if the agent could accurately categorize this partner as, for example, a likely cooperator or defector. This inference could be based on perceivable features of the partner in question, such as their age, personality, or social status. Supervised learning from labeled examples is the study of algorithms which learn predictive models by generalizing from past observations (Bishop 2006; Hastie et al. 2001). In our game-theoretic example, past observations refer to previously encountered partners, along with a label categorizing their observed behavior. A predictive model accurately categorizes new partners when only the feature values, such as those mentioned above, are known.

Supervised-learning problems are formalized by first defining an input space $X$ of feature vectors used by the agent to encode observations, and an output space $Y$ specifying the structure of the labels being predicted. Observations are drawn from the product space $Z = X \times Y$. Categorization tasks are those where $Y$ is a set of category labels. Regression tasks are those where $Y$ is some range of numeric values. An environment which is unknown to the agent determines the underlying functional relationship between inputs and outputs, and is given by a joint probability distribution $\mu(\mathbf{x}, y)$. Given a multiset of $r$ labeled observations $S = \{\langle \mathbf{x}_i, y_i \rangle\}_{i=1}^{r} \in Z^r$, the agent uses a learning algorithm to select a hypothesis which represents an informed guess at which systematic pattern best explains the functional relationship between inputs and outputs, not just between the observed examples, but in general.

Hypotheses describe what is systematic in the observations, and allow the agent to make predictions about novel objects, the problem of guessing *y* given only **x**. The hypothesis space $\mathcal{H}$ defines the set of hypotheses the agent can select from, and plays a critical role in its ability to generalize accurately from experience. The learning algorithm is also critical because it determines which hypothesis is selected for a given series of observations. A learning algorithm, *L*, implements a mapping,

$$L : \bigcup_{r \geq 1} Z^r \mapsto \mathcal{H}, \tag{5.1}$$

from multisets of observations to hypotheses. It will prove useful to view the hypothesis space as a model $f_\theta(\mathbf{x})$, indexed by the parameters $\theta$. Selecting a hypothesis is the process of estimating the parameters $\theta$ from observations.

### 5.2.2 Error-inducing risk and uncertainty

An omniscient observer has full knowledge of the environment, which is assumed to be fully specified by the data-generating distribution $\mu(\mathbf{x}, y)$. A hypothesis induced by an agent can also be seen as a joint probability distribution over the observations,[1] and denoted $\sigma(\mathbf{z})$. The discrepancy between these two distributions, $\mu$ and $\sigma$, will be defined as their Kullback-Leibler divergence, given by

$$D(\mu \| \sigma) = \sum_{\mathbf{z} \in Z} \mu(\mathbf{z}) \log_2 (\mu(\mathbf{z}) / \sigma(\mathbf{z})). \tag{5.2}$$

The Kullback-Leibler divergence is perhaps best understood from a coding perspective, where a probability distribution over the observation space implies an optimal coding scheme which assigns a code of length $log_2(1/p)$ bits to an observation occurring with probability *p*. The greater the probability of the observation, the shorter its code. The Kullback-Leibler divergence measures the number of additional bits required to code events governed by $\sigma$ when using the code for $\mu$ (Cover and Thomas 1991). When the distributions are identical, $D(\mu \| \sigma) = 0$. We will use this measure of divergence to arrive at an abstract categorization of three basic forms of discrepancy, referred to as stochasticity, underspecification, and misspecification. These discrepancies can exist independently, and their potential combinations are depicted in Figure 5.1. Both the organism and the theorist attempting to understand the organism face these uncertainties.

---

[1] Here we assume that the agent learns a generative model from the observations, one which models both $\Pr(y|\mathbf{x})$ and $\Pr(\mathbf{x})$ (see Bishop 2006, for further discussion).

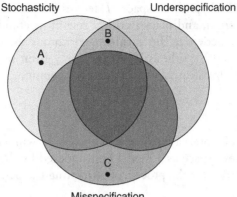

Figure 5.1. Possible combinations of the three basic forms of discrepancy: stochasticity, underspecification, and misspecification. The discrepancies are independent of each other, and all combinations are possible. Point A corresponds to the problem of predicting the outcomes of a fair die. Point B corresponds to the problem of predicting the relative frequency of red and black balls in an urn. Point C corresponds to the problem of predicting the toss of a loaded die which always falls on 4, using a misspecified hypothesis space which allows predicting only 1 or 6.

### Stochasticity

Consider an agent with certain knowledge of $\mu$, such that $\mathcal{H} = \{\sigma\}$ and $D(\mu\|\sigma)=0$. This agent will make error-free predictions if outputs are always a deterministic function of the inputs. If the predictive distribution $\Pr(y|\mathbf{x})$ is stochastic, such that for at least one input $\mathbf{x}$, the output is nonunique, then error will result. Stochasticity is the most basic form of discrepancy between agent and environment. It arises due to external randomness and cannot be eliminated through further observation, or by designing the organism differently. For example, even though an agent knows that a die is fair, it will make errors when predicting rolls of the die (Figure 5.1, point A). In short, even when granted full causal knowledge of the process governing observations – such as knowledge of physical probabilities (Giere 1999), propensities (Popper 1959), or a priori probabilities (Knight 1921) – the agent will still make errors when predicting events under conditions of stochasticity.

### Underspecification

Discrepancies due to underspecification exist when the number of observations underdetermines the choice of hypothesis, and is too small to reliably converge on the best model $\sigma^*$ in some $|\mathcal{H}| > 1$, where

$$\sigma^* = \min_{\sigma \in \mathcal{H}} D(\mu \| \sigma). \tag{5.3}$$

At one extreme, in the complete absence of observations, the agent faces what Knight (1921) refers to as uncertainty. For Knight, an agent may know that an urn contains red and black balls (knowledge coded implicitly in the hypothesis space) but the probability of choosing a black ball from the urn is uncertain if no other information is available (Figure 5.1, point B). As the number of observations increases from zero, the agent moves from an uncertainty situation to a risk situation, and can begin measuring what Knight terms statistical probabilities. Underspecification, in Knightian terms, will therefore include situations of both risk and uncertainty.

### Misspecification

Discrepancies due to misspecification result from the agent's inability to model the data-generating distribution exactly, such that $D(\mu \| \sigma^*) > 0$. Misspecification can occur in the absence of stochasticity and underspecification. To take a simple example, an agent entertaining two hypotheses, $\mathcal{H} = \{$"die always falls on 1", "die always falls on 6"$\}$, will make errors when predicting a loaded die which always falls on 4 (Figure 5.1, point C). Such errors are unrelated to stochasticity and underspecification.

### Nonstationarity

So far we have implicitly assumed that the probability distribution on the observation space – which determines what we observe – is the same as the distribution which determines the accuracy of our inferences. Situations involving nonstationarity are those where these two distributions differ. We will assume that nonstationary problems represent a special case of misspecification, since any temporal dependency on the distribution generating observations, $\mu_1(\mathbf{x},y)$, and the distribution which determines the accuracy of predictions, $\mu_2(\mathbf{x},y)$, can be reformalized by considering a single distribution, $\mu_t(\mathbf{x},y)$, parameterized by time $t$. The class of nonstationary problems nevertheless includes forms of discrepancy worthy of study in their own right, and poses problems of great practical significance (e.g. Quiñonero-Candela et al. 2009; Hand 2006).

The generating distribution $\mu(\mathbf{x},y)$ when rewritten $\Pr(y|\mathbf{x})\Pr(\mathbf{x})$ highlights two broad categories of nonstationarity. First, covariate shift occurs when $\Pr(\mathbf{x})$ changes (Shimodaira 2000). Second, what we term predictive shift occurs when $\Pr(y|\mathbf{x})$ changes. Figure 5.2 illustrates how the two forms of

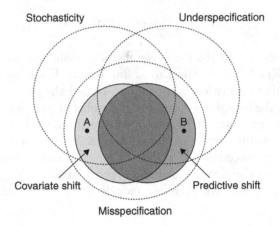

Figure 5.2. Nonstationarity as a form of misspecification. Covariate shift refers
to a change in the probability distribution over inputs. Predictive shift refers to a
change in conditional probability of the outputs given the inputs. These two forms of
nonstationarity can occur independently. Point A, discussed in the main text, refers to
an example of sample selection bias, where partners in a game-theoretic setting may be
of similar age. Point B is an example of predictive shift, also discussed in the main text,
where potential partners may change their strategy.

nonstationarity can exist independently and always represent a form of misspecification. An example of covariate shift is sample selection bias, which
occurs when a potentially unknown process probabilistically rejects observations (Heckman 1979). For example, in our game-theoretical setting, the
partners an agent experiences may be restricted to be those of a similar age
(Figure 5.2, point A). Here, the predictive distribution remains constant but
our observations of it will be skewed. Predictive shift could occur if potential partners became aware of an agent's decision process, and adjusted their
behavior to exploit it. Here, the functional relationship between observable
features and outcomes may change (Figure 5.2, point B).

### 5.2.3  Further uncertainties

These basic uncertainty categories refer to discrepancies between an
environment, experienced through a series of observations, and abstract
properties of the hypothesis space. A full categorization of real-world
uncertainties is infeasible and unbounded, and some basis for separat-
ing exogenous from endogenous uncertainty is always necessary. For
example, all biological systems operating at temperatures above 0 kelvin

must be robust to the uncertainties arising from thermal noise, but we will abstract from this problem even though it imposes significant constraints on functional design (Wagner 2005). Perhaps the most significant source of error-inducing discrepancy to be sidestepped in the coming discussion will be the costs associated with errors. In particular, we assume that the prices paid for incorrect predictions are given by standard loss functions, such as zero–one loss for classification problems, and squared difference for regression problems. Organisms in natural environments must contend with events incurring potentially very different costs, such as the devastating consequences of the highly improbable (Taleb 2009; Bookstaber and Langsam 1985).

In addition to the basic uncertainty categories defined above, one must also consider uncertainty surrounding the algorithm $L$ used to select hypotheses. What we will term computational uncertainty arises from both fundamental and idiosyncratic barriers which constrain the processing of observations. Such constraints introduce error-inducing discrepancies, just like the basic categories discussed above. Misspecification, for example, arises because finite observations will tend to undermine the selection of the most predictive hypothesis. In the same way, resource restrictions on $L$ can limit the ability of the organism to arrive at a predictive model. Constraints on $L$ range from primitive considerations of computability and intractability, to additional, organism-specific biological and computational resource limitations. The former apply to all computational agents (Hopcroft and Ullman 1979; Papadimitriou 1995). The precise nature of the latter will depend on the agent, and include constraints such as limitations on working memory. Unlike many sources of uncertainty, computational uncertainty poses a fundamental problem for all agents in all contexts, with the only exception being omnipotent, and therefore fictitious, agents.

## 5.3 FROM ERROR TO OPTIMALITY

An omniscient observer knows the true state of nature. A rational actor achieves the best possible response relative to a set of assumptions about the true state of nature. Consequently, our degree of faith in a rational actor model should be in proportion to our knowledge of the problem. When facing large worlds, we should question both the validity of our models, and our use of the terms "rational" and "optimal." These issues focus the remainder of discussion onto the following question: Under conditions of stochasticity, underspecification, and misspecification, what does it mean for an organism to be rational, and respond optimally? Answering this question will first require taking a closer look at the relationship between the organism, the environment, and error.

### 5.3.1 Analyzing and categorizing error

Relative to the generating distribution $\mu$, the hypothesis $f(\mathbf{x})$ chosen by the agent after observing a sample of observations $S$ is likely to incur some degree of error. To simplify matters, we will focus on the problem of regression where error is commonly defined as the squared difference between the true and predicted value. The lowest achievable error is incurred by the function $g(\mathbf{x})$, defined as the conditional expectation $E[y|\mathbf{x}] = \int yp(y|\mathbf{x})dy$ of the predictive distribution $\Pr(y|\mathbf{x})$. Relative to the generating distribution, the total error incurred by $f(\mathbf{x})$ is referred to as the expected loss,

$$\text{expected loss} = \int \{f(\mathbf{x}) - g(\mathbf{x})\}^2 \, p(\mathbf{x})\,d\mathbf{x} + \int \{g(\mathbf{x}) - y\}^2 \mu(y, \mathbf{x})\,d\mathbf{x}\,dy. \quad (5.4)$$

Notice that the first term of this expression is dependent on the chosen hypothesis, $f(\mathbf{x})$, but the second term is not, and therefore it cannot be reduced, whatever our choice of $f(\mathbf{x})$. This second term expresses the error incurred by the agent when always selecting $f(\mathbf{x}) = g(\mathbf{x})$. This part of the error corresponds directly to error arising from stochasticity, which is commonly referred to as irreducible error or noise.

*Controllable components of error*
The first term of Equation 5.4 measures that part of the error we can reduce through the appropriate design of the agent, since it depends on the policy for selecting $f(\mathbf{x})$ from observations. Adopting a frequentist perspective, and imagining that the tape of experience were replayed several times, we will use a standard technique in statistical learning theory for decomposing the controllable error into two components, referred to as bias and variance (O'Sullivan 1986; Geman et al. 1992; Bishop 2006; Hastie et al. 2001). For each replay of the tape of experience, a potentially different sample $S = \{\langle \mathbf{x}_i, y_i \rangle\}_{i=1}^{r}$ will be observed, where $S$ is a multiset of $r$ observations sampled from $\mu(y, \mathbf{x})$. Now, replaying the tape of experience $k$ times yields $k$ of these multisets, given by the ensemble $\mathcal{S} = \{S^{(1)}, S^{(2)}, ..., S^{(k)}\}$.

Rather than the error incurred by the agent after observing a specific sample, we would like to know about its mean error, relative to the ensemble of hypotheses selected by the agent's learning algorithm for each member of $S$. For a given $\mathbf{x}$, the expectation of the first term of Equation 5.4 for an ensemble of samples of size $k$ is $\mathbb{E}_S\left[\{f(\mathbf{x}) - g(\mathbf{x})\}^2\right]$. Integrating over $\mathbf{x}$, this expectation is the sum of two terms, the first of which is bias,

$$(\text{bias})^2 = \int \{\mathbb{E}_S[f(\mathbf{x})] - g(\mathbf{x})\}^2 \, p(\mathbf{x}) dx, \qquad (5.5)$$

which is the squared difference between mean prediction made by the functions induced from the ensemble, and the true function $g(\mathbf{x})$, over all inputs. The second term is the variance, given by

$$\text{variance} = \int \mathbb{E}_S\left[\{f(\mathbf{x}) - \mathbb{E}_S[f(x)]\}^2\right] p(\mathbf{x}) dx, \qquad (5.6)$$

which is the squared difference between the mean prediction of the ensemble, and the predictions of the individual functions induced for each member of the ensemble. Collecting the terms for bias, variance, and irreducible error, the expected loss given by Equation 5.4 can be summarized as

$$\text{Total error} = (\text{bias})^2 + \text{variance} + \text{irreducible error}. \qquad (5.7)$$

By decomposing the error in this way, we have a more precise understanding of the effects of misspecification, underspecification, and stochasticity. The bias and variance incurred by the algorithm will always be a function of the sample size, $r$, and the properties of the generating distribution. In general, variance decreases as a function of $r$, the size of the observed sample. Variance occurs when changes to the sample lead to changes in the error of the induced hypothesis. Limiting variance requires reducing the sensitivity of the learning algorithm to the effects of resampling. Generally speaking, bias occurs due to an inability of the algorithm's hypothesis space to model the generating distribution.

*The bias/variance dilemma*
Keeping in mind that the generating distribution is unknown to the agent, considerations of bias and variance highlight a fundamental problem in inductive inference known as the bias/variance dilemma. At one extreme, the agent's learning algorithm could express a wild guess by ignoring the observations altogether, by always selecting the same hypothesis. This approach guarantees zero variance, but can lead to high

bias unless the guess turns out to be correct, or close to correct. At the other extreme, the agent's learning algorithm could hedge its bets, let the observations speak for themselves, and select a hypothesis from a highly flexible model space capable of approximating any function. This policy could in principle guarantee zero bias, but usually at the expense of high variance since the flexibility of the hypothesis space is likely to lead to an oversensitivity to the vagaries of particular samples. The bias/variance dilemma arises because methods for minimizing variance tend to increase bias, and methods for minimizing bias tend to increase variance. The two need to be balanced, a process which should be guided by knowledge of the task at hand.

### 5.3.2 The relationship between bias, variance, and optimality

What is the optimal response to the bias/variance dilemma? Zero variance and zero bias will be achieved by a learning algorithm which always induces the same hypothesis, the conditional expectation of the predictive distribution $g(\mathbf{x})$. Recall, though, that we are interested in cases where the generating distribution is not known with certainty. Taking a Bayesian perspective, the assumptions made about the generating distribution, which are required to generalize from data in any way, are coded in the structure of the hypothesis space and the prior. Optimality is always defined relative to these assumptions. If we knew the generating distribution, then the hypothesis space should contain a single hypothesis, $g(\mathbf{x})$. If we knew the functional form of the generating distribution was, for example, linear, then the inference problem reduces to estimating the parameters of the linear model. Optimality is not about the presence or absence of error, but the degree of error relative to a set of assumptions. The optimal algorithm incurs bias and variance, like any other algorithm. The label "optimal" simply marks out a particular algorithm as incurring the least error among all algorithms sharing the same hypothesis space and prior.

### 5.4 FROM SMALL WORLDS TO LARGE WORLDS

What is a small world problem? The most restrictive definition of a small world problem is one where the optimal response is certain, and uncontroversial. For problems of inductive inference, optimal responses are usually framed in terms of Bayesian statistics. These optimality results rely on two conditions being met. First, the structure and properties of the problem are known a priori, such that the appropriate hypothesis space and prior are known with certainty, rather than inferred from observations. Second, the structure of the hypothesis space and prior permit calculation of the posterior probability using exact methods. The first condition settles any dispute over the external validity of the model – the degree to which the properties of the model match those of the system being modeled. The second condition settles any dispute over the internal validity of the model – the degree to which the inferences made using the model are rationally justified, assuming that the model is correct, and has external validity. Obviously, very few problems of interest to the cognitive, behavioral, and social sciences can be said to satisfy the first condition. Similarly, many problems, when formalized in these terms, require use of inexact, approximate Bayesian methods. However, this definition of a small world guarantees that statements of rationality and optimality provide exactly what they advertise. This guarantee is possible because all relevant aspects of the problem, and solution, are certain. In practice, though, rational actor models are not absolute statements but statements about optimality relative to some set of assumptions.

The notion of a large world becomes relevant as soon as the structure and properties of the problem are inferred from observations, and issues of misspecification, underspecification, nonstationarity, and all the other categories of uncertainty we discussed begin to impact on the external validity of the model. In addition, issues of analytic tractability impacting on the process of determining the posterior distribution can also introduce uncertainty. Thus, even in cases where we can guarantee an appropriate formalization of the problem, the analytic and computational demands of performing rational calculation also need to be considered, as they can lead to additional error. Now, all good theorists know that "a map *is not* the territory" (Korzybski 1958, p. 750) and all models are "wrong." Given this, why question the objective of rational actor models? After all, if your underlying assumptions are openly stated, then everyone is free to assess the implications of your model as they see fit. Our point is that increasing uncertainty and ignorance should at some point lead us to question the notions of rationality and optimality. When factors known to compromise the internal and external validity of the model are at play, declarations of rationality and optimality become less and less meaningful. Next, we show that the use of rational actor models and optimality results in the analysis of large world problems is a choice, not a necessity (Brighton and Olsson 2009).

### 5.4.1 Distinguishing relative from absolute function

In developing a rational actor model one is attempting to answer the question of how to make optimal inferences, predictions, given assumptions about the properties of the task at hand. When these properties are uncertain, how else can functioning be assessed? We will contrast the pursuit of rational actor models with an alternative. Borrowing the terminology of Breiman (2001), rational actor models rely on data modeling, an approach used routinely in statistics where one assumes an underlying statistical model, and then uses observations to estimate the model parameters. We will adopt an approach referred to by Breiman as algorithmic modeling, where the underlying statistical model is treated as unknown (or nonexistent), and the estimated predictive accuracies of alternative statistical models, or learning algorithms, are then compared in an attempt to learn about the problem. These two approaches are both mute on the problem of assessing the rationality of organisms. To illustrate their use in examining the ability of organisms to function in uncertain environments we will use two fictional modelers, named Jones and Smith.

Modeler Jones follows a methodology known as the rational analysis of cognition, an influential approach inspired by optimality modeling in biology (Anderson 1991; Oaksford and Chater 1998, 2007). A rational analysis typically proceeds by inferring from observations of the task environment an appropriate observation space, hypothesis space, and prior over the hypothesis space. These components define a probabilistic model of the task environment. Jones then derives an optimal response function which specifies the inferences that a rational, usually Bayesian observer would make in response to further observations of the environment. The rationality of the agent in question is then assessed by comparing its behavioral responses with those of the rational model. If the rational model provides a close fit to the agent's responses, Jones argues that the agent makes optimal inferences.

Modeler Smith is interested in exactly the same problem, begins by observing the task environment, and also proposes an appropriate observation space. Crucially though, Smith *refrains* from making an inference about an optimal response, an inference which requires postulating an hypothesis space and generating distribution. Instead, Smith estimates the predictive accuracy of potentially several statistical models *relative to observations* of the environment. Predictive accuracy is often measured using cross-validation, a technique we illustrate below (see Rissanen

1986, for an alternative relativistic approach which stresses compression rather than prediction). This comparison informs the question of which mechanistic design features prove functional, rather than optimal, in the task environment. The ability of each mechanism to describe the agent's behavior is an additional question, guided by functional analysis, but assessed by experimentation (Brighton and Gigerenzer 2011). In contrast to the approach of Jones, Smith refrains from making an inference about the generating distribution, conducts a competitive test of rival process-level theories instead of seeking a single rational actor model, and therefore sees success as something always relative to alternative models, rather than being defined by the benchmark of a single rational actor model.

### 5.4.2 Relative functioning in large worlds

What insights can Smith's relativist approach offer? As an illustration, consider the problem of how a retail marketing executive might distinguish between active and inactive customers based on their purchasing history. Current thinking on how to best address this problem centers on stochastic customer-base-analysis models, such as the fairly sophisticated Pareto/negative binomial distribution model. Following modeler Jones' approach, this model defines rationally justified inferences about customers, given an assumed underlying probability distribution. In the spirit of modeler Smith's approach, Wübben and Wangenheim (2008) compared the performance of several models, including the Pareto/negative binomial distribution model, and found that the best-performing model was a simple hiatus rule which predicts inactivity if a customer has not made a purchase in the preceding nine months (and for one problem, six months). To take another example, consider the problem of searching literature databases, where the task is to order a large number of articles so that the most relevant ones appear at the top of the list. When examining this problem, Lee et al. (2002) constructed a "rational" Bayesian model and found that in comparison to several competing models, a simple one-reason heuristic proved superior.

What can these, and similar findings, tell us? Models implementing a "rational" response to an underlying distribution which is assumed, but inaccurate, can result in performance inferior to that of a model which makes no explicit attempt to model an underlying distribution, or conduct any form of rational calculation. Many people find results like these surprising, but modeler Jones may argue that they say nothing about rational actor models in general, they merely supply an argument against poorly designed rational actor models. Jones is correct, but misses the point. In a large world, one cannot identify a rational model, or an optimal response, with any confidence. Rather, the "rational" response is to search for better and better models. In this sense, the pursuit of rational actor models for large worlds can be counterproductive. First, one cannot assume that rational calculation brings us closer and closer to effective responses to uncertainty. Second, the alternative is not silence, but an examination of alternative models, particularly those which have been successful in the past. Perhaps the worst case scenario is that Jones' approach focuses functional analyses onto small world problems which bear little relation to the large world problems of interest. Next, we focus on the first issue, which concerns common intuitions about rational calculation.

## 5.5 HOW TO CONFRONT LARGE WORLDS

Mathematical objects such as generating distributions and rational actors could be viewed as misnomers when set against the uncertainty surrounding natural environments and the constraints impacting on computational actors. From a rational actor perspective, these complicating factors limit the opportunities for rigorous functional analysis. But rather than defining limits on rational analysis, these complexities highlight a need for alternative approaches (e.g. McNamara and Houston 2009). The relativist approach taken by our fictional modeler Smith is standard practice when comparing the predictive accuracy of learning algorithms and cognitive models on real-world datasets (e.g. Perlich et al. 2003; Brighton 2006; Chater et al. 2003; Czerlinski et al. 1999; Gigerenzer and Goldstein 1996). Datasets are simply collections of observations relating covariates to a dependent variable. For example, the daily stock prices over the last year, test results for a population of patients undergoing treatment, or a series of utterances and their associated meanings. Datasets are usually samples from an unknown and potentially unknowable distribution, often suffer from uncertainty arising from noisy or missing information, and may represent a snapshot of a nonstationary problem. In much the same way, an organism functioning in an environment, and a scientist attempting to explain this organism's functioning, both face problems shot through with uncertainty. The question of which strategy is optimal under these circumstances cannot be answered. But we can study why some strategies repeatedly prove more effective than others.

### 5.5.1 Making inferences with simple heuristics

The study of simple heuristics illustrates several crucial issues. Simple heuristics are cognitive process models which ignore information (Gigerenzer et al. 1999; Gigerenzer and Brighton 2009; Gigerenzer and Gaissmaier 2011). A central question in the study of heuristics is understanding when and why they outperform alternative cognitive models. For many, the idea that ignoring information can prove functional is counterintuitive. Consider, for instance, the simple heuristic, take-the-best (Gigerenzer and Goldstein 1996). Take-the-best is a cognitive process model describing how people decide which of two objects has a greater value on some criterion of interest, such as the price of two houses, or the maximum speed of two vehicles. If the criterion values are unknown, environmental cues such as the presence of a swimming pool or whether the vehicle has a jet engine or not, can be used to make an inference. More precisely, take-the-best has three steps:

(1) Search rule: Search through cues in order of their validity.
(2) Stopping rule: Stop on finding the first cue that discriminates between the objects (i.e. cue values are 1 and 0).
(3) Decision rule: Infer that the object with the positive cue value (1) has the higher criterion value.

In short, take-the-best searches for the first cue which discriminates between the objects, and then makes an inference using this cue alone. For example, a house with a swimming pool is likely to be more expensive than a house without. All other cues are ignored. Specifically, take-the-best searches for cues in order of their validity, which is a measure of how accurately each cue has made inferences on previous comparisons. This measure is simple in comparison with the measures used by most models, as it ignores any dependencies between cues. That is, take-the-best orders cues by assuming that the predictive ability of a cue can be determined independently of the value of the other cues.

Take-the-best is a linear model.[2] Furthermore, it deviates from common statistical intuition, and the methods employed by commonly assumed models of cognitive processing. For example, the Gauss-Markov theorem proves that among all unbiased linear models, the least-squares estimate incurs the lowest variance (e.g. Fox 1997). Assuming that the

---

[2] This is true at an outcome level, but take-the-best differs at a process level from the standard linear models, such as linear regression.

problem in question is linear, this tells us that a logistic regression model, for example, should incur lower error than take-the-best. How can we put this statement to the test? A classic task in the study of simple heuristics is the problem of inferring which of two German cities has a greater population. The dataset in question describes the eighty-three largest German cities using nine binary cues which detail properties of the cities, such as whether or not a city has a university or an intercity train station, or is located in Germany's industrial belt. After observing some sample of these cities, and estimating the model parameters from this sample, the accuracy of competing algorithms, like take-the-best and logistic regression, can be measured. Accuracy here refers to whether or not larger cities are correctly inferred as being larger. Crucially, the accuracy of these inferences is measured by considering only novel pairs of cities, comparisons between cities which were *not* used to estimate the model parameters. This process is repeated for different samples, and is termed cross-validation (Stone 1974).

Figure 5.3 plots the mean predictive accuracy of take-the-best as a function of the size of sample used to estimate the cue validities. In the same plot, we report the predictive accuracy of logistic regression. Means are reported, which are taken with respect to 5,000 random partitions of the dataset into training and testing sets. Now, why, in contrast to our interpretation of the Gauss-Markov theorem, does take-the-best outperform logistic regression so significantly? The Gauss-Markov theorem, despite often being interpreted as statistical justification for the widespread use of the least-squares linear model, is a small world theorem. The theorem holds when fitting, rather than predicting, data. As soon as one makes predictions about unseen data, the effects of underspecification and misspecification can change matters significantly.

A common response to this result, and similar findings for many other datasets (e.g. Brighton 2006; Czerlinski et al. 1999), is that logistic regression is a weak competitor, and more sophisticated methods designed specifically for avoiding the problem of overfitting will outperform take-the-best (e.g. Chater et al. 2003). In response, Figure 5.3 also compares the performance of take-the-best with a support vector machine, one of the most advanced methods in statistical pattern recognition (Vapnik 1995). Again, take-the-best outperforms the support vector machine over a significant portion of the learning curve. Notice that we have compared three models without postulating, or making assumptions about, a generating distribution determining the relationship between properties of German cities and their populations. Instead, we have respected the fact

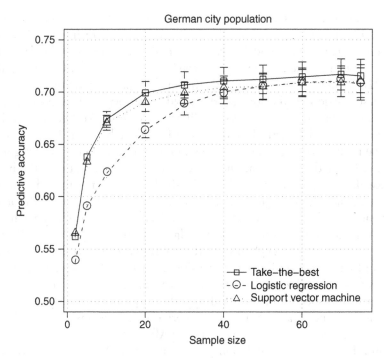

Figure 5.3. Can ignoring information be beneficial? For the German city population task, the predictive accuracy of the simple heuristic take-the-best is plotted as a function of the size of the sample used to estimate the model parameters. Take-the-best, despite ignoring conditional dependencies between cues and using only one cue to make an inference, outperforms both logistic regression and a support vector machine. Mean predictive accuracies are reported for 5,000 random partitions of the dataset. Error bars show the variance.

that "a set of data is one thing and a mathematical object, such as a distribution, is quite another, not only different in degree but different in kind" (Rissanen 1986, p. 395), and measured the relative ability of the models to make accurate inferences relative to the observations. How, though, are results like these related to the question of decision making in large worlds?

### 5.5.2 The bias/variance dilemma in large worlds

Recall from our discussion of the bias/variance dilemma that the problem of making accurate predictions can be decomposed into two subproblems: the problem of reducing variance, and the problem of reducing bias. Variance arises due to underspecification, a lack of observations, while bias tends to arise as result of misspecification, an inherent inability on the part of the model to capture predictive patterns in the data. In their influential presentation of the bias/variance dilemma, Geman et al. (1992) discussed how restricting the hypothesis space can reduce variance. If these restrictions do not introduce additional bias with respect to the problem at hand, prediction error is likely to decrease. Another approach, taken by several statistical models, is to introduce bias which *is* relevant to the problem at hand, under the assumption that it will be outweighed by a greater reduction in variance. Recently, we showed that take-the-best succeeds by exploiting this trick; take-the-best's performance advantage arises exclusively through a reduction in variance by ignoring conditional dependencies between cues (Brighton and Gigerenzer 2007; Gigerenzer and Brighton 2009). The success of ridge regression and the naïve Bayes classifier can also be explained in these terms (e.g. Hoerl and Kennard 2000; Domingos and Pazzani 1997; Friedman 1997; Hastie et al. 2001).

From a rational actor perspective, where one conducts rational calculation with respect to an assumed hypothesis space and prior probability distribution, how can the success of alternative algorithms, like the naïve methods mentioned above, be explained? After all, why should deviations from a full Bayesian calculation prove functional? Advocates of rational actor models will likely argue that the success of any model can be explained using rational principles. For example, one can ask under which conditions take-the-best is optimal, and thereby frame take-the-best as a rational actor model under these conditions. Thus, it is trivially true that the success of heuristics, or any other model, in no way challenges the small world study of rationality. Two points are crucial here. First, while seeking a rational explanation for apparently nonrational processes can certainly yield great insight, this task is far from being a trivial exercise, and will often achieve only a limited, approximate explanation (e.g. Sanborn et al. 2010). The optimality conditions of the naïve Bayes classifier, for example, are only partially known, despite sustained study and a clear Bayesian interpretation (e.g. Domingos and Pazzani 1997; Kuncheva, 2006). Nevertheless, the naïve Bayes classifier appears in the top-ten list of data mining algorithms, and is used routinely in uncertain contexts, those where the modeler is largely ignorant of the generating distribution (Wu et al. 2007). Second, this issue has little or no impact on the question of what strategies prove functional in a large world. If the underlying structure of the problem is unknown, then optimality results offer little help. In large worlds, all we can do is estimate the predictive accuracy of a range of methods, such as those algorithms which have performed well in the past. The outcome of such an exploration will never be a proof of optimality, nor a state of understanding which tells us what is "rational." In large worlds, we have no choice but to abandon the objective of an optimal response, and instead search for improved understanding. Possessing a rational interpretation of the algorithms used to conduct this search changes nothing.

### 5.5.3 Understanding large worlds

We defined small worlds as presenting problems where all relevant factors are certain and their formalization yields to feasible rational calculation. Uncertainty enters the picture when the characteristics of the problem must be inferred from observations, or when the complexity of the problem, once formalized, renders rational calculation infeasible. Faced with some degree of ignorance, we have two options. First, we can develop a rational actor model by making the assumptions necessary to recast the problem as a certain, small world problem. Second, we can acknowledge that an optimal response to an inherently uncertain problem is a fiction, and likely to obscure our understanding (Klein 2001). Instead, we should aim to examine the relative ability of competing models to explain the observations. The best model we can find is neither optimal nor rational, merely functional.

Put in these terms, at least two factors need to be considered in any proposed definition of a large world. First, one must gauge the impact of the basic uncertainty categories we introduced earlier, such as underspecification, misspecification, and nonstationarity. From our perspective, real-world problems of theoretical significance tend not to be small world problems. We assume a large world by default, which is why we study heuristics rather than rational actor models. Our goal is to understand how, in mechanistic terms, organisms cope with uncertainty in the large worlds to which they are adapted. The second factor to consider is the analytic tractability of performing rational calculation. Thus, it would be an oversight not to mention here that exact computation of the posterior is only possible for certain priors; brute force calculation is only tractable for very restricted hypothesis spaces; and computationally tractable algorithms rely almost exclusively on inexact methods which have optimality guarantees which are asymptotic in nature (e.g. Bishop 2006; MacKay 2003).

Rational principles themselves can be disputed, even in small worlds. For classification tasks, the problem of assigning a class $y \in Y$ to a novel observation $\mathbf{x}$, the gold standard is the Bayes optimal classifier, which given a sample of observations $S$, a hypothesis space $\mathcal{H}$, and prior $\Pr(\mathcal{H})$, assigns a class $y$ to an observation by

$$y = \max_{y_i \in Y} \sum_{H_j \in \mathcal{H}} \Pr(y_i \mid \mathbf{x}, H_j)\Pr(\mathcal{H}_j \mid S). \tag{5.8}$$

The class $y$ assigned to $\mathbf{x}$ is calculated by considering the predictions of all hypotheses in $\mathcal{H}$, with each prediction weighted by the posterior probability of the hypothesis. Despite Equation 5.8 being widely regarded as defining an optimal response for classification problems, Grünwald and Langford (2007) have proven its suboptimality under certain forms of misspecification (see also Diaconis and Freedman 1986). More broadly, rationality principles used in the study of inductive inference are necessarily based on statistical assumptions. For instance, Bayesian rationality and algorithmic information theory are closely related due to the relationship between probability and coding (which we drew on when explaining Kullback-Leibler divergence), but they can differ both in theory (Vitányi and Li 2000; Grünwald 2005) and practice (Kearns et al. 1997; Pitt et al. 2002). Subtleties such as these highlight that "rationality principles are invented rather than discovered" (Binmore 2009, p. 2) and we should question the view that there is one rationality that follows directly from the laws of probability theory, logic, or some other calculus.

In our introduction, we used the example of the aircraft manufacturer, and argued that no engineer can hope to arrive at an accurate model of relevant mechanical, atmospheric, and human factors which determine the safety of a flight control system. Furthermore, even if this were possible, the analytic task of finding an optimal response to an appropriate criterion, even if such a criterion exists in any meaningful sense, is beyond the abilities of any expert. We view the significant questions facing the study of inductive inference as having more in common with this large world engineering problem than with the development of rational actor models for small world problems. The study of large world inference problems, like the study of flight control systems, is the search for mechanisms which are more and more robust to uncertainty, but also the study of when and why these mechanisms prove robust (Kitano 2004; Wagner 2005; Hammerstein et al. 2006). This second task can certainly profit from the study of rational actor models and optimality modeling. But this fact in no way questions or undermines the substantive, empirical issue of understanding robust responses to large worlds. Rational inquiry should not be confused with rational actor models. The former is a commitment to sound science, but the latter is a commitment to a particular approach to modeling.

## 5.6 ARE RATIONAL ACTOR MODELS "RATIONAL" OUTSIDE SMALL WORLDS?

Savage introduced the notion of a small world when questioning the validity of Bayesian decision theory under conditions of uncertainty. We see Savage's question as applying more broadly, beyond decision theory, and impacting directly on the two key questions which guide the study of inductive inference in humans and other animals. First, how, in mechanistic terms, do organisms arrive at inductive inferences? Second, what is the relationship between the behavior we observe in organisms, and the optimal behavior as defined by a rational actor model? Now, in attempting to answer the second question, do we run the danger of committing a Type III error, and finding the right answer to the wrong question? In contrast to the idea that humans are biased, error-prone users of dumb heuristics, there is a growing tendency to view humans as astonishingly well-adapted to an uncertain world, as evidenced by the ability of humans to handle uncertainty in ways which exceed even the most sophisticated human-engineered machinery (Geman et al. 1992; Poggio and Smale 2003; Tenenbaum et al. 2006). In this sense, evolved biological organisms can be seen as existence proofs of adaptive responses to large, uncertain worlds. This perspective makes the first question, the question of uncovering the underlying cognitive and perceptual mechanisms of humans and other animals, the key question.

Rather than hallmarks of advanced understanding, we have argued that the concepts of rationality and optimality have the potential to be counterproductive when examining large worlds. Without doubt, Bayesian decision theory and Bayesian statistics, to take two examples, inform the question of what is rational in a small world. The crux of the issue is then to assess the dangers of using the same concepts to understand large worlds. While it is possible to analyze uncertain worlds by introducing assumptions which discharge uncertainty, and recast the problem as a small world problem, we have argued for an alternative. The alternative is to follow a relativist approach which makes no attempt to identify an optimal, rational response, but instead aims to incrementally improve understanding by identifying models with greater and greater predictive accuracy. Crucially, this approach dispenses with the need to make the assumptions necessary to recast the problem as a small world problem. For some problems, we have to accept that our uncertainty and ignorance make finding optimal, rational responses a meaningless endeavor. This fact, though, does not make the study of uncertainty meaningless.

## REFERENCES

Anderson, J. R. (1991). Is human cognition adaptive? *Behavioral and Brain Sciences*, 14: 471–517.

Binmore, K. (2009). *Rational Decisions*. Princeton University Press.

Bishop, C. M. (2006). *Pattern Recognition and Machine Learning*. Springer, New York.

Bookstaber, R. and Langsam, J. (1985). On the optimality of coarse behavior rules. *Journal of Theoretical Biology*, 116: 161–193.

Breiman, L. (2001). Statistical modeling: The two cultures. *Statistical Science*, 16(3): 199–231.

Brighton, H. (2006). Robust inference with simple cognitive models. In Lebiere, C. and Wray, R., editors, *Between a Rock and a Hard Place: Cognitive Science Principles Meet AI-Hard Problems* (AAAI Technical Report SS-02-06), pages 189–211. AAAI Press, Menlo Park, CA.

Brighton, H. and Gigerenzer, G. (2007). Bayesian brains and cognitive mechanisms: Harmony or dissonance? In Chater, N. and Oaksford, M., editors, *The Probabilistic Mind: Prospects for Bayesian Cognitive Science*, pages 1179–1191. Cambridge University Press.

(2011). Towards competitive instead of biased testing of heuristics: A reply to Hilbig and Richter (2011). *Topics in Cognitive Science*, 3: 197–205.

Brighton, H. and Olsson, H. (2009). Identifying the optimal response is not a necessary step toward explaining function. *Behavioral and Brain Sciences*, 32: 85–86.

Chater, N., Oaksford, M., Nakisa, R., and Redington, M. (2003). Fast, frugal, and rational: How rational norms explain behavior. *Organizational Behavior and Human Decision Processes*, 90: 63–86.

Cover, T. M. and Thomas, J. A. (1991). *Elements of Information Theory*. Wiley, New York, NY.

Czerlinski, J., Gigerenzer, G., and Goldstein, D. G. (1999). How good are simple heuristics? In Gigerenzer, G., Todd, P. M., and ABC Research Group, *Simple Heuristics That Make Us Smart*, pages 119–140. Oxford University Press.

Diaconis, P. and Freedman, D. (1986). On the consistency of Bayes estimates. *The Annals of Statistics*, 14(1): 1–26.

Domingos, P. and Pazzani, M. (1997). On the optimality of the simple Bayesian classifier under zero-one loss. *Machine Learning*, 29: 103–130.

Fox, J. (1997). *Applied Regression Analysis, Linear Models, and Related Methods*. Sage Publications, Thousand Oaks, CA.

Friedman, J. H. (1997). On bias, variance, 0/1-loss, and the curse-of-dimensionality. *Data Mining and Knowledge Discovery*, 1: 55–77.

Geman, S., Bienenstock, E., and Doursat, R. (1992). Neural networks and the bias/variance dilemma. *Neural Computation*, 4: 1–58.

Giere, R. G. (1999). *Science without Laws*. University of Chicago Press.

Gigerenzer, G. and Brighton, H. (2009). *Homo heuristicus:* Why biased minds make better inferences. *Topics in Cognitive Science*, 1: 107–143.

Gigerenzer, G. and Gaissmaier, W. (2011). Heuristic decision making. *Annual Review of Psychology*, 62: 451–482.

Gigerenzer, G. and Goldstein, D. G. (1996). Reasoning the fast and frugal way: Models of bounded rationality. *Psychological Review*, 103(4): 650–669.

Gigerenzer, G., Todd, P. M., and ABC Research Group (1999). *Simple Heuristics That Make Us Smart*. Oxford University Press.

Grünwald, P. (2005). Minimum description length tutorial. In Grünwald, P., Myung, I. J., and Pitt, M. A., editors, *Advances in Minimum Description Length*, pages 23–79. MIT Press, Cambridge, MA.

Grünwald, P. and Langford, J. (2007). Suboptimal behavior of Bayes and MDL in classification under misspecification. *Machine Learning*, 66: 119–149.

Hammerstein, P., Hagen, E. H., Herz, A. V. M., and Herzel, H. (2006). Robustness: A key to evolutionary design. *Biological Theory*, 1(1): 90–93.

Hand, D. J. (2006). Classifier technology and the illusion of progress. *Statistical Science*, 21: 1–14.

Hastie, T., Tibshirani, R., and Friedman, J. (2001). *The Elements of Statistical Learning: Data Mining, Inference, and Prediction*. Springer, New York.

Heckman, J. J. (1979). Sample selection bias as a specification error. *Econometrica*, 47(1): 153–161.

Hoerl, A. E. and Kennard, R. W. (2000). Ridge regression: Biased estimation for nonorthogonal problems. *Technometrics*, 42: 80–86.

Hopcroft, J. E. and Ullman, J. D. (1979). *Introduction to Automata Theory, Languages and Computation*. Addison-Wesley, Reading, MA.

Kearns, M., Mansour, Y., Ng, A. Y., and Ron, D. (1997). An experimental and theoretical comparison of model selection methods. *Machine Learning*, 27: 7–50.

Kitano, H. (2004). Biological robustness. *Nature Reviews Genetics*, 5: 826–837.

Klein, G. (2001). The fiction of optimization. In Gigerenzer, G. and Selten, R., editors, *Bounded Rationality: The Adaptive Toolbox*, pages 103–121. MIT Press, Cambridge, MA.

Knight, F. H. (1921). *Risk, Uncertainty, and Profit*. Houghton Mifflin Co., Boston, MA.

Korzybski, A. (1958). *Science and Sanity*, 4th edition. International Non-Aristotelian Library Publishing Co., Lakeville, CT.

Kuncheva, L. I. (2006). On the optimality of Naïve Bayes with dependent binary features. *Pattern Recognition Letters*, 27: 830–837.

Lee, M. D., Loughlin, N., and Lundberg, I. B. (2002). Applying one reason decision-making: The prioritization of literature searches. *Australian Journal of Psychology*, 54: 137–143.

MacKay, D. J. (2003). *Information Theory, Inference, and Learning Algorithms.* Cambridge University Press.

McNamara, J. M. and Houston, A. I. (2009). Integrating function and mechanism. *Trends in Ecology and Evolution*, 24: 670–674.

Oaksford, M. and Chater, N., editors (1998). *Rational Models of Cognition.* Oxford University Press.

(2007). *Bayesian Rationality.* Oxford University Press.

O'Sullivan, F. (1986). A statistical perspective on ill-posed inverse problems. *Statistical Science*, 1: 502–518.

Papadimitriou, C. H. (1995). *Computational Complexity.* Addison-Wesley, Menlo Park, CA.

Perlich, C., Provost, F., and Simonoff, J. S. (2003). Tree induction vs. logistic regression: A learning curve analysis. *Journal of Machine Learning Research*, 4: 211–255.

Pitt, M. A., Myung, I. J., and Zhang, S. (2002). Toward a method of selecting among computational models of cognition. *Psychological Review*, 109(3): 472–491.

Poggio, T. and Smale, S. (2003). The mathematics of learning: Dealing with data. *Notices of the American Mathematical Society*, 50: 537–544.

Popper, K. R. (1959). The propensity interpretation of probability. *British Journal of the Philosophy of Science*, 10: 25–42.

Quiñonero-Candela, J., Sugiyama, M., Schwaighofer, A., and Lawrence, N. D., editors (2009). *Dataset Shift in Machine Learning.* MIT Press, Cambridge, MA.

Rissanen, J. (1986). Stochastic complexity and statistical inference. *Lecture Notes in Control and Information Sciences*, 83: 393–407.

Sanborn, A. N., Griffiths, T. L., and Navarro, D. J. (2010). Rational approximations to rational models: Alternative algorithms for category learning. *Psychological Review*, 117(4): 1144–1167.

Savage, L. J. (1954). *The Foundations of Statistics.* Wiley, New York.

Shimodaira, H. (2000). Improving predictive inference under covariate shift by weighting the log-likelihood function. *Journal of Statistical Planning and Inference*, 90: 227–244.

Stone, M. (1974). Cross-validatory choice and assessment of statistical predictions. *Journal of the Royal Statistical Society B*, 36: 111–147.

Taleb, N. N. (2009). Errors, robustness, and the fourth quadrant. *International Journal of Forecasting*, 25: 744–759.

Tenenbaum, J. B., Griffiths, T. L., and Kemp, C. (2006). Theory-based Bayesian models of inductive learning and reasoning. *Trends in Cognitive Sciences*, 10(7): 309–318.

Vapnik, V. (1995). *The Nature of Statistical Learning Theory.* Springer-Verlag, New York.

Vitányi, P. M. B. and Li, M. (2000). Minimum description length induction, Bayesianism, and Kolmogorov complexity. *IEEE Transactions on Information Theory*, 46(2): 446–464.

Wagner, A. (2005). *Robustness and Evolvability in Living Systems*. Princeton University Press.

Wu, X., Kumar, V., Quinlan, J. R., Ghosh, J., Yang, Q., Motoda, H., McLachlan, G. J., Ng, A., Liu, B., Yu, P. S., Zhou, Z.-H., Steinbach, M., Hand, D. J., and Steinberg, D. (2007). Top 10 algorithms in data mining. *Knowledge and Information Systems*, 14: 1–37.

Wübben, M. and Wangenheim, F. von (2008). Instant customer base analysis: Managerial heuristics often get it right. *Journal of Marketing*, 72: 82–93.

# CHAPTER 6

# *Pull, push or both?*

## Indirect evolution in economics and beyond

### Siegfried Berninghaus, Werner Güth and Hartmut Kliemt

## 6.1 INTRODUCTION

The issue addressed in this chapter can be introduced in the following terms:

When there is a formal relationship between two interpretations, like the rationalistic conscious maximization and evolutionary interpretations of Nash equilibrium, then one can say, 'Oh, this is an accident, these things really have nothing to do with each other.' Or, one can look for a deeper meaning. In my view, it's a mistake to write off the relationship as an accident. (Aumann 1998, p. 192)

We follow Aumann in searching for a 'deeper meaning' though what we find will not necessarily be the meaning that he imagines. In doing so we try to explore the relationship between an 'eductive'[1] and an 'evolutionary' approach to interactive choice making, in a philosophical rather than an analytical way.[2] These two perspectives may also be referred to as *pull* (intentional, planned) versus *push* (evolutionary, adaptive); the contrast between them has been around since the beginnings of social theory (Section 6.2).[3] They can be combined in an *indirect evolutionary approach* providing 'a deeper meaning' to the relationship (Section 6.3) and embedded into a broader conceptual framework of layered approaches

---

The authors gratefully acknowledge the editors' detailed criticism and advice. Their many constructive suggestions helped to improve this chapter considerably. Of course, the conventional disclaimer applies.

[1] *Eductive* is Binmore's (1987) term for conscious maximization.

[2] For a summary of the exercises, in which at least one of the present authors has been involved, see 'Our Exercises on Indirect Evolution' (Max Planck Institute of Economics n.d.). For other exercises, see Samuelson (2001). This is, of course, a very incomplete selection of the papers on indirect evolution.

[3] 'Push and pull' has been used as a description of the critical impact of pre-Newtonian modern mechanics. Early modern theorists would reject explanations and law-like regularities which could not be spelled out in terms of push and pull as mysterious. We require here too that explanations be spelled out, but we are, of course, aware that physics since Newton admitted such mysterious claims as forces operating at a distance.

(Section 6.4). After our constructive analysis of the paradigm case of the basic binary game of trust, we turn to a methodological criticism of some alternative accounts of the matter (Section 6.5). The chapter ends with general concluding remarks (Section 6.6).

## 6.2 A LITTLE HISTORY OF THOUGHT

Evolutionary thought is not a modern invention. In a passage of his *Physics* that we moderns, with hindsight, came to regard as significant, Aristotle states:

A difficulty presents itself: why should not nature work, not for the sake of something, nor because it is better so, but just as the sky rains, not in order to make the corn grow, but of necessity? What is drawn up must cool, and what has been cooled must become water and descend, the result of this being that the corn grows. ... Why then should it not be the same with the parts in nature, e.g. that our teeth should come up of necessity – the front teeth sharp, fitted for tearing, the molars broad and useful for grinding down the food – since they did not arise for this end, but it was merely a coincident result; and so with all other parts in which we suppose that there is purpose? Wherever then all the parts came about just what they would have been if they had come to be for an end, such things survived, being organized spontaneously in a fitting way; whereas those which grew otherwise perished and continue to perish, as Empedocles says his 'man-faced ox-progeny' did. (*Physics* 198b; i.e. book II, part 8, Aristotle n.d.)

'Yet', adds Aristotle two sentences later, 'it is impossible that this should be the true view'. As an empiricist he had to pass a negative verdict on evolutionary explanations, since the empirical evidence and broader background theories of the time counted very strongly in favour of 'teleology'.

We use the term 'teleology' here and subsequently as referring to explanations of events in terms of purposeful behaviour. Such explanations can be perfectly in line with the assumption that purposeful behaviour is itself subject to behavioural laws (of nature). The laws of nature are operating blindly or without a purpose at each point in time and the fact that entities pursuing ends emerged in the course of time can itself be explained as a result of processes of 'Darwinian evolution' (which as a whole lacks a 'telos' and is therefore classified as non-teleological here).

The potential explanation of order in terms of 'blind' and 'mindless' evolution that Aristotle cites was at his time merely an unconvincing thought experiment. Until the end of the seventeenth century modern thinkers, notably Spinoza (see Spinoza 1992, bk 1, prop. 36, appendix), who postulated that whatever happens occurs out of pure 'necessity' and without a 'telos' were still lacking a convincing paradigm case in which the emergence of a complex order could be understood without relying on some kind of plan. Only after theorists like Mandeville, Ferguson, Hume and Smith had developed and applied the idea of 'results of human action that are not consequences of human planning or design' (see Schneider 1967)[4] was the way paved for Darwin's and Wallace's idea of natural selection among living beings. Their ideas found expression in social Darwinism, popular thought and then back into economics. Schumpeter is a case in point:

The assumption that conduct is prompt and rational is in all cases a fiction. But it proves sufficiently near to reality, if things have time to hammer logic into men. Where this has happened, and within the limits it has happened, one may rest content with this fiction and build theories upon it ... and we can depend upon it that the peasant sells his calf just as cunningly and egoistically as the stock exchange member his portfolio of shares. But this holds good only where precedents without number have formed conduct through decades and, in fundamentals, through hundreds and thousands of years, and have eliminated unadapted behavior. Outside of these limits our fiction loses its closeness to reality. (Schumpeter 1959, p. 80)

Regrettably, many later economists who have defended the model of homo economicus along evolutionary lines disregarded 'Schumpeter's caveat' that outside the limits of repeated interaction the 'fiction loses its closeness to reality'. Eager to defend far-reaching economic rationality assumptions, they gave in to the temptation of applying the argument from adaptation as though it leads to optimization across the board.

It is clearly illegitimate to assume that the circumstances Schumpeter invokes to explain the emergence of rational economic behaviour apply universally in social reality. Evolutionary selection may work both for and against the ability to be 'pulled' by perceived opportunities. To get a full understanding of the phenomena of human interaction we have to look at it as both 'pushed' and 'pulled'. It is worth developing a systematic rather than merely metaphorical account of how the push and pull perspectives hang together. In Section 6.3 we introduce a simple example of such integrated modelling (6.3.1 and 6.3.2), then indicate how the basic idea may be generalized (6.3.3) and extended to aspects of games beyond preferences (6.3.4).

---

[4] It should be noted that Ferguson attributes his own formula to the seventeenth-century thinker Cardinal de Retz (*Memoirs*). Yet as Russell Hardin (2007, p. 48 n. 56) states it cannot be found there. In any event, older inspirations notwithstanding, it took the eighteenth-century moralists to provide a detailed and convincing story.

## 6.3 INDIRECT EVOLUTION AS AN INTEGRATED APPROACH

So-called prosocial behaviour among human individuals is not exceptional. If it were not for economics, most theorists would not regard it as a remarkable phenomenon at all. True to the ironic saying that everything has been said before but not by us, we economists join the chorus belatedly, but with our own voice. In particular, experimental economists have documented the prevalence of certain restrictions on local maximization of an observable maximand (like monetary payoff) in ways that we find convincing (e.g. Plott and Smith 2008).[5] But we have yet to fully come to terms with these facts and to integrate them into our models successfully. The following discussion of a simple case of human interaction involving trust will illustrate the point.

### 6.3.1 Direct evolution in the 'trust interaction'

Consider first a trust interaction described in terms of objective payoffs. Modelling such an interaction from the point of view of an external onlooker rather than a participant, we are led to the general scheme depicted in Figure 6.1.

If agent $A$ chooses $D_A$, or 'defection', the payoffs will be (0, 0). The choice of $C_A$ by $A$ is classified as a 'co-operative' move. When $B$ comes to move, he can 'co-operate' by choosing $C_B$, leading to (1, 1), or can 'defect' by choosing $D_B$ leading to $(a, b)$.[6] Whether the individuals in first-mover or '$A$'-roles play $C_A$ is determined by a fixed behavioural program; similarly for their partners in '$B$'- or second-mover roles. Both players do what they do not on the basis of a conscious model of the situation (pull); they do it because they are 'programmed' that way (push).

Let us assume that the interaction of Figure 6.1 is repeated indefinitely among randomly matched pairs of individuals picked from an infinite population. In this single-population model all individuals are assigned to both roles with probability ½. As long as the population share $p$ of $C_B$-individuals is sufficiently large, i.e. as long as $p + (1 - p)a > 0$,

---

[5] The fact that experimental economic results also indicate that prosocial behaviour is often subject to erosion in repeated interactions does not imply that such behaviour may be neglected. When it comes to basic issues of organizing social interaction *both* the presence of prosocial behaviour and its tendency to diminish unless re-enforced must be taken into account.

[6] Though the payoffs are 'objective' rather than 'subjective' we may still assume that the unit of measurement can be appropriately adjusted such as to fix the units at '1' and '0' respectively – all that matters is relative success as measured in objective terms.

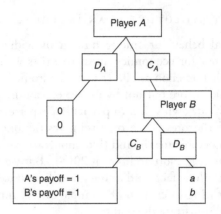

Figure 6.1. Simple trust interaction with *b > 1 > 0 > a.*

those who play $C_A$ in role 'A' fare better than those who do not. If playing $C_A$ is a disposition that is passed on to the individuals in the next generation, then the share of $C_A$-individuals should increase. However, $D_B$-individuals do better than $C_B$-individuals across the board and therefore their share will increase in the presence of $C_A$-individuals. The greater the number of $C_A$-individuals the better $D_B$-individuals fare relative to $C_B$-individuals. Regardless of the initial distribution, $D_B$-individuals will drive out $C_B$-individuals. Therefore eventually $p + (1 − p)a < 0$ must hold, and the $D_A$-types drive out $C_A$-types. In the end a monomorphic population characterized by the behavioural program $(D_A, D_B)$ will emerge. Since it cannot be invaded by $C$-types it is the only evolutionarily stable state.

The preceding discussion is an outline of an elementary direct evolutionary story as told from the viewpoint of an external observer. It is not mentally represented by the phenotypes involved in the plot. Once mental models of the situation are present and proximate factors become operative within the evolutionary process, we need to pay attention to what the actors 'know'. Interactions of entities thereby become 'games' between 'actors' in the proper sense of that term. There is now 'pull' as well as 'push'. The latter might be reconstructed in eductive terms and appended to the evolutionary description.

### 6.3.2 Indirect evolution of preferences and the game of trust

The trust interaction becomes a game by assuming that the objective payoffs, which determine relative success in the evolutionary process, are mapped onto subjective ones, representing the proximate motivations. For convenience we generally assume that the actors are motivated subjectively by the objective payoffs.[7]

Although for simplicity we use the same numerical values to represent the objective and the subjective payoffs, there are always two distinct evaluations involved (one representing what happens at the evolutionary level, the other what goes on on the motivational level).

We assume that the participants in the interaction think about it in terms that can be represented by the model of Figure 6.1. They know that they are involved in the interaction as depicted by the game model and the semantic rules of interpreting such a model. All know that they, as well as all others, know and understand the game model, know that this is known, know that it is known that it is known, and so on: common knowledge of the rules of the game prevails among participants.[8]

*Imperfect 'type information'*

The preceding model becomes more interesting if information about the players' types is imperfect. In Figure 6.2 the horizontal line connecting the two decision nodes of player $A$ indicates that he has to choose $C$ without knowing which type of individual will act in the role of the second mover (i.e. the two connected decision nodes of player $A$ are in one information set). We are then led to the game of Figure 6.2 (in which subscripts to the co-operative and non-co-operative moves '$C$' and '$D$' of players $A$ and $B$ have been suppressed.)

Note that the payoffs in Figure 6.2 are subjective, representing preferences with $b > 1 > b^*$. The subjective payoffs coincide numerically with the objective ones, since we assume that individuals in both roles are entirely motivated by objective payoff, except for the payoff $b^*$. This discrepancy is due to the fact that two types of players exist in the population. On the one hand, we have a population share $p$ of trustworthy $b^*$-types, who reward first-mover trust rather than exploit it, $(1 > b^*)$. They do so regardless of the higher objective payoff of '1' (in Figure 6.2) that they would receive if they were to use their $D$-strategy against their

---

[7] The way in which we deviate from the conventional assumption that 'utility is linear in money' will be indicated and explained in some detail in our discussion of Figure 6.2, below.

[8] In experiments conditions of common knowledge are often created by informing participants 'in public', i.e. in an event in which all participate simultaneously in ways that make them aware of their common presence. We make the otherwise rather strong common knowledge assumption without further ado, since the eductive perspective that we intend to discuss in relation to the evolutionary one is based on it anyway. In another context it would be necessary – and possible – to give a cognitive psychology account of the knowledge conditions.

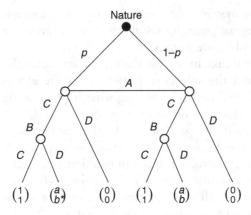

Figure 6.2. Trust with private type information.

subjective inclination not to do so. The $b$-type individuals are motivated by objective payoffs across the board. They subjectively prefer to exploit first-mover trust, receiving the higher objective payoff of $b$ instead of the payoff of 1 from rewarding trust, ($b > 1$).

To have a well-defined game, we assume that the shares $p$ of trustworthy $b^*$-types and $1 - p$ of untrustworthy $b$-types are common knowledge and determine consistent prior beliefs.

Like the preceding model of direct evolution, the model of indirect evolution used here assumes a single population of actors who are randomly assigned a role: the first-mover or $A$-role and the second-mover or $B$-role. In this game, using terms like 'trust' for playing $C$ as first mover and 'exploitation' for playing $D$ as second mover makes sense, since now intentions and deliberate choices are involved.

Though in the last resort it may be misleading to conceive of trust solely in terms of rational betting,[9] this conception is a useful first approximation. An individual in the $A$-role will have a reason to bet on the 'trustworthiness' of individual $B$ iff $p + (1 - p)a > 0 \Leftrightarrow p > a/(a - 1)$. Since $p$ is assumed to be common knowledge, all individuals will behave in the same way in first-mover roles. Cognitive expectations concerning the behaviour of other individuals rather than their own behavioural type determine the actions of first movers. Rationality and forward looking choice rather than a fixed behavioural program (like a disposition to choose 'C' in first-mover

[9] For a game-theoretically informed critique of identifying trusting with rational betting, see Lahno 2002.

roles) will determine the outcome of the game. All rational actors will switch to showing 'no-trust', $D$, in the first-mover role once $p$ becomes too low, $p < a/(a - 1)$, while in the first-mover role all – independently of their type – will show 'trust', $C$, if $p$ is large enough, $p > a/(a - 1)$.

As long as the commonly known $p$ is large enough, all participants in the game will choose to 'trust'. The untrustworthy who use $D$ in second-mover roles will flourish and outcompete the trustworthy who use $C$. The share $p$ of trustworthy individuals will shrink since – as evaluated in objective terms – the $D$-behaviour of the untrustworthy after a $C$-move of the first mover will yield more than the trustworthy $C$-behaviour in the second-mover role.

Once the odds are against betting on trustworthiness, because $p$ has become too low, evolutionary pressure against the trustworthy disposition will cease to exist unless we assume that actors in first-mover roles occasionally make mistakes and trust. Whenever this happens an untrustworthy second mover will fare better than a trustworthy second mover. In view of this, we may assume that evolution at the slow, mistake-driven pace will reduce the share of the trustworthy even below the minimum population share for which trust is a rational choice.

In summary, we can conclude that whenever information about players' types is completely private in evolutionary games of trust like the one sketched above, only individuals who are subjectively motivated to maximize objective payoff on every round of play in the role of the trustee will survive. Moreover, even without the argument of a mistake-driven evolution beyond the minimum population share for which showing trust is the rational bet, we can at least state that under the condition of completely private type information, rewarding and trusting behaviour cannot evolve.

### Piercing the veil of privacy of type information

Assume now that individuals can either signal or detect trustworthiness, or both, such that they are able to detect at some cost and with some reliability the type of their co-player. It is quite easy to show that for sufficiently low cost $C$ and sufficiently high reliability (which we take as given), there is a generic range of population compositions in which the trustworthy will outcompete the untrustworthy.

In Figure 6.3 the shape of the triangle is determined by the reliability and the costs of the detection technology. The reliability is determined by the probabilities that a trustworthy $b^*$-type and an untrustworthy $b$-type are each signalled as such (also fixing the smaller complementary probabilities of receiving a mistaken signal). In view of the reliability of the detection technology and the payoff parameters of the game, the interval of population shares $p$ for which it is rational for a player to follow the signal can be determined by a little algebra (see for details Güth and Kliemt 2000). For the present more philosophical purposes, some graphical illustrations should suffice.

Continuing with the discussion of Figure 6.3, consider the baseline of the triangle – $C = 0$ – which illustrates the relevant interval for which it is advantageous for actors in first-mover roles to follow the signal once it has been received.[10] For alternative cost-level $C$, the iso-cost-line of points $(p, C)$ falling into the interior of the triangle indicates the population compositions $p$ for which – at cost $C > 0$ – it is worthwhile to acquire the detection technology. Once the costs, $C$, become prohibitively high – i.e. $C > C''$ – the interval in which it is worthwhile to acquire the technology for receiving the signal becomes empty.

If the technology for type detection is not prohibitively costly but rather available at cost $C'$ all rational individuals will acquire it for $p \in (\pi, \Pi)$ as indicated in Figure 6.3. The parameter $s = a/(a - 1) \in (\pi, \Pi)$ indicates the threshold at which individuals without specific type information would shift their behavioural gears. For $s > p$, trust is a bad bet. Individuals who invested in the detection technology would, however, trust after receiving a signal that their specific partner is trustworthy – provided that following the signal is rational. For $p$ between $\pi$ and $s$, individuals try to single out trustworthy individuals by means of the detection technology. Beyond $s$, i.e. $s < p < \Pi$, the desire to avoid those who are not worth trusting is the reason for investing in the technology until $p$ becomes so small that unconditional trust without any sorting attempt is the superior strategy.

Figure 6.3 presents combinations $(p, C)$ of population compositions $p$ and costs $C$ that must be incurred in discriminating between different types in second-mover roles. For initial values of $p_0$ for which the combination $(p_0, C)$ is located outside the triangle, the technology is too costly to be used by first movers. There are either too few or too many trustworthy individuals around to make it worthwhile to pay the price for having access to the technology.

For any combination $(p_0, C)$ within the triangle, discrimination between trustworthy and non-trustworthy individuals in second-mover

---

[10] Treating $C > 0$, after investing it, as a sunk cost is tantamount to treating it as if $C = 0$ were to apply.

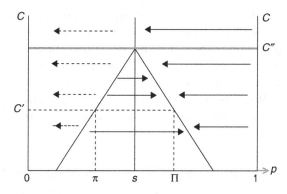

Figure 6.3. Population dynamics with imperfect type information.
*Note:* For details see Berninghaus et al. 2003.

roles is sufficiently reliable to induce first movers to bear the cost $C$. This renders trustworthiness – as emerging due to subjective modifications of objective payoffs – objectively superior to non-trustworthiness. The share $p$ of trustworthy types will increase within the interior of the triangle.

More specifically, start with an initial population composition $p_0 > \pi$ and cost $C'$. Drawing the horizontal line from $(\pi, C')$ by varying only $p$ until the line hits the right edge of the triangle at $(\Pi, C')$ we know the evolutionarily stable bimorphism (characterized by $\Pi$) that will be reached for all $p_0 > \pi$ if costs are $C'$.

For costs $C$ beyond $C''$ the technology will never be used. Type information remains private and only the monomorphic population characterized by $p = 0$ can be evolutionarily stable for $C > C''$. For positive cost $C \le C''$ and for any given reliability of detection at that cost, the stable composition $p = 0$ will be reached as well, if the initial population composition $p$ is to the left of the triangle.

It is obvious that only the trustworthy would survive if perfect individual type information became common knowledge at no cost. In that case $b^*$-types would attract trust from all individuals in first-mover roles and fare objectively better than $b$-types. Even though, objectively speaking, $b$-types would receive higher payoffs *if* trusted, no first mover would trust them. Due to perfect discrimination, the untrustworthy never intentionally get a chance to exploit first movers while the trustworthy are always offered the chance to reward trust. The trustworthy will always reap the co-operative payoff of 1, while the untrustworthy end up with 0. As measured in objective payoffs, the untrustworthy end up worse off than the trustworthy. Under perfect type discrimination, the population should eventually be composed exclusively of trustworthy types.[11]

The preceding discussion focuses on a paradigmatic example illustrating conditions under which teleological decision making makes a difference. The question naturally arises whether more general conclusions can be drawn. We turn to this question next, before considering how the indirect evolutionary methodology can be applied to issues other than preference evolution.

### 6.3.3 Some general characteristics of indirect preference evolution

*The ultimate criterion of success need not be represented in preferences*
At the proximate level of behavioural explanation, adaptive behaviour can be seen as motivated by rules and norms rather than by the pursuit of the objectively better payoff in each and every situation. Although the motive and proximate explanation of the behaviour may not involve maximizing objective payoffs, the behaviour may nevertheless tend to increase those payoffs over the long term. Rules that are objectively payoff maximizing can be followed for all sorts of reasons. The actor, though ultimately maximizing objective payoff, is not necessarily a maximizer of objective payoff at the proximate level.

Under suitable informational conditions, there exist evolutionarily stable population compositions with a non-negative share of trustworthy individuals who are rule-bound, rather than unconstrained, maximizers. Straightforward maximization of objective payoffs is not the only strategy that can survive over the long term. This raises the general question of the relation between objective payoffs and subjective motivations. More technically: what are the conditions under which evolutionarily stable preferences can be represented by utility functions that are positive linear transformations of the objective payoff applying locally (i.e. to each decision situation taken separately)?[12]

*Conditions favouring the coincidence of preference and ultimate success criteria*
The preceding example makes it clear that whether type information is private or not may have a decisive influence on the survival prospects of

---

[11] It may be noted in passing that these individuals are of a specific type rather than merely behaving as if they were of that type.
[12] These would be the cases in which the conventional assumption that utility is 'linear in money' or linear in 'substantive payoff' applies. In these cases the objective payoff function is one of the class of utility representations of 'subjective preferences'.

types. If type information is completely private, then only individuals that are maximizers of objective payoff in every instance can survive in the evolutionary game of trust.

It can also be shown that proximate profit-maximizing behaviour will survive if individual actions are strategically insignificant in interactions with other individuals. If actions do not affect others there will be no reactions from other individuals. The anticipation of reactions cannot then provide a sufficiently strong incentive to pursue material payoff maximization. In particular, in a perfectly competitive market in which no individual can strategically influence others, only straightforward payoff maximization can survive.

It can indeed be shown that under certain circumstances private-type information and insignificance of individual actions are sufficient for profit maximization to survive (see Güth and Peleg 2001). The circumstances that must be fulfilled are of a rather technical nature, e.g. continuous games must have an interior solution amounting to a strict Nash equilibrium such that derivatives fulfil appropriate conditions for local maxima. If that is so, then each of the two conditions is on its own (and thus also jointly) sufficient for objective payoff maximization to be the only proximate motive and mode of behaviour that can be evolutionarily stable.

It should be observed that the circumstances supporting the preceding results are quite special. Under different circumstances, there can be an ecological niche for motives other than payoff maximization. In particular, the ability to detect the types of other individuals, and the ability to deceive them about one's own type, will both change the character of the interaction.[13]

### 6.3.4 Indirect evolutionary studies extended

#### Co-evolution

Consider again the trust interaction of Figure 6.2. Assume not only that the preference type of the actor in the second-mover role can vary, but that there are now also informed and uninformed types who play in the first-mover role. There will be an interaction between the two kinds of type variation. For example, if co-player types are all alike, then discriminatory efforts will be rendered futile. This will put an evolutionary premium on *not* developing the faculty to discriminate (assuming the faculty

---

[13] They form an essential part of the rules of each game, emerging as part of the evolutionary process.

Table 6.1. *Player types in simple games of trust.*

| One population model strategies for the simple game of trust | | First-mover role | |
| --- | --- | --- | --- |
| | | *I*(nformed) | *U*(ninformed) |
| Second-mover role | *t* (trustworthy) | *t, I* | *t, U* |
| | *n* (non-trustworthy) | *n, I* | *n, U* |

is costly). On the other hand, if the population of co-players is heterogeneous, then developing the ability to discriminate may be worthwhile in objective terms, even if it is costly.

If we use '*t*' to indicate trustworthiness and '*n*' non-trustworthiness in second-mover roles, '*I*' for informed and '*U*' for uninformed player types in first-mover roles, we get the range of possible combinations depicted in Table 6.1 (again in a 'single-population' setting in which all actors can play in both roles).

In the game of trust, the presence of first movers who are able to discriminate between trustworthy and untrustworthy actors in the second-mover role will change the prospects of trustworthy and untrustworthy individuals, respectively. At the same time, the value of the ability to discriminate between different types of co-players depends on the population composition. If, for example, all potential co-players are trustworthy the ability to discriminate does not lead to better proximate decisions, and therefore will not pay off ultimately in the objective terms that determine evolutionary success.

Analysing the co-evolution of information status ('*U*', '*I*') in first-mover and preferences ('*t*', '*n*') in second-mover roles is technically tricky (see Güth et al. 2000). Eventually, once the fraction of informed types $q$ exceeds some threshold, the composition of the population cycles in closed orbits around a rest point $(q^*, p^*)$. Cycling in the two dimensional space of all $(p, q)$ with $0 \leq p \leq 1$ and $0 \leq q \leq 1$, where $p$ is the fraction of trustworthy types and $q$ the fraction of informed types is illustrated in Figure 6.4.

*Extending the basic ideas*

The basic ideas can be extended to settings in which, for instance, cost parameters are dependent on the population composition and thus become endogenous to the evolutionary process. If the costs of discriminating

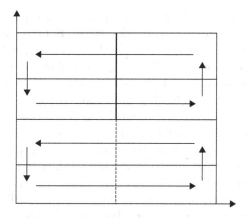

Figure 6.4. Cycling of type compositions in $(p, q)$-space.

trustworthy types are dependent on how many trustworthy types are around, we have to consider $C(p)$ rather than a constant C. Again, the objective success of types is ultimately decisive for their survival, but proximately they decide on the basis of the cost parameters perceived by them in each period.

In the same vein, the co-evolution of preferences and social or legal institutions can be studied. For instance, if there are courts to enforce certain norms of trustworthy behaviour, then there will be incentives even for the opportunists to behave as if they were trustworthy, which may in turn reduce the incentive to become informed about one's co-player's type in interactions involving trust. However, if the demand for trustworthiness declines, its value for those who are endowed with it may decline as well. So it is conceivable that courts and legal institutions could crowd out trustworthiness (see also Frey 1997).

How the prevalence of trustworthy types in the population might affect court behaviour is another issue that can be studied on the basis of a co-evolutionary model. In a population in which trustworthiness as an inclination to act according to rules prevails, it is intuitively plausible that courts will work better and might even strengthen the presence of trust-worthiness. Similarly, the co-evolution of preferences and of sanctioning behaviour without specialized court institutions may be studied within an indirect evolutionary approach (see Brennan et al. 2003 and Güth and Ockenfels 2005).

In the preceding and other examples, the conventional rational choice model based on forward-looking optimization is used to derive the results of games on the proximate level of gameplay. This has the advantage of staying close to conventional economic analysis. But there is no obstacle in principle to substituting the conventional rational choice model by some model of bounded rationality. The specific type of model used to represent intentional forward-looking choice is less important than the fact that the analysis includes both an objective and a subjective dimension.

It goes without saying that embedding some concept of bounded rationality – or, perhaps, full-fledged cognitive psychology on the proximate level – into the overall evolutionary process is complicated (see for such attempts Güth and Kliemt 1998, and Güth et al. 2005). In exploring these possibilities we should, however, be aware that bounded rationality is no less a teleological concept than perfect rationality, and as such distinct from evolutionary concepts. The *push* sphere in which the anticipated future exerts its motivational influence is different from the *push* sphere in which 'blind' forces operate without anticipation of future consequences. This distinction is crucial to our modelling strategy.

### 6.4 THEORETICAL LAYERS AND THE INTERACTION OF LAYERS

Very general insights of social theory tend to be elementary. The preceding discussion suggests three such elementary, general points: (1) An adequate theoretical understanding of social reality requires layers of teleological and non-teleological theories. (2) As in natural science, layers cannot be logically reduced to one another, yet have to fulfil the requirement of general coherence between the separate layers.[14] (3) The 'interaction' between layers cannot be neglected.

#### 6.4.1 Teleology within evolution and evolution within teleology

The simplest illustration of the layered approach arises if we consider merely pairs of theoretical layers. There are four possible combinations of teleology and evolution, or of push and pull in a two-layered setting (Table 6.2).

Cell 1 of Table 6.2 relates to a two-layered process of choice making in which both layers are teleological. For example, at the higher rational level, people purposefully agree on the rules of a poker game to be played, and

---

[14] See Stöckler 1991 for a useful brief account of reductionism as implied but not explicitly discussed here.

Table 6.2. *Push and pull combinations at different explanatory levels.*

| Lower level/higher level | Teleological in terms of choice making | Evolutionary in terms of selective adaptation |
|---|---|---|
| Teleological in terms of choice making | 1 | 3 |
| Evolutionary in terms of selective adaptation | 2 | 4 |

then subsequently make their lower-level choices within the rules agreed on. Typically some kind of rational decision theory is applied to describe decision making on both levels (e.g. as in Buchanan and Tullock 1962; Brennan and Buchanan 1985). Assuming that informationally closed substructures emerge, we could treat the whole process as one game in which different subgames can be reached. The anticipated solutions of the subgames – the lower levels – determine what will be the rational course of action on the higher level.[15] Once its initial node has been reached, a subgame is closed with respect to external causal influences and closed with respect to the future. Causal feedback loops, as in the indirect evolutionary approach, do not exist.

Cell 2 refers to the case of predicting in adaptive terms the results of a game chosen on the higher level. Somebody might use game-theoretic analysis and strategies of rational choice on the higher level. At that level, the choices are explained as being teleological and forward-looking. The outcomes of substructures chosen on the higher level are predicted in natural terms, perhaps by an evolutionary process more narrowly conceived.

An example arises when we consider a person like Ulysses, who anticipates his own compulsive future behaviour and responds by engaging in rational self-management (see, for instance, Ainslee 2002; Elster 1979; Schelling 1984). Another example is a theorist who deliberately relies on genetic algorithms to find solutions for problems that he will then use on a higher level. Still another example would be a so-called *market maker,*

---

[15] As stated in Binmore (1987), much depends on whether we believe that deviations from eductive rationality are explained by uncorrelated trembles or not. If the background ontology ascribed to the decision makers themselves is based on the distinction between 'choice making' by a doer (who from his own internal point of view generates causal effects, see Pearl 2000) and behaviour of an organism as explained by psychological (or physical) laws then layers of reality are naturally involved; see, for a more extended discussion from our point of view, Güth and Kliemt 2007.

who creates a game whose specific outcomes he cannot anticipate but foresees that it is generally advantageous to let the play of the game run its course.

The examples of the indirect evolutionary approach presented in Section 6.3, fit in cell 3. The higher level is evolutionary and the lower level teleological. The higher-order process is not to be explained in terms of choice making. The lower-order process is teleological and explained either in psychological or in eductive terms.

Cell 4 is of interest as well. Here we are dealing with a two-layered evolutionary process. There is a longer time horizon of ultimate adaptation and a shorter horizon of proximate adaptation. Both are evolutionary without any reference to teleology. The involvement of two layers becomes significant due to the 'speed differentials' and the feedback loop.[16] But let us look more closely at the relationship between the two layers in the specific model of the 'indirect evolutionary approach to trust' discussed above.

### 6.4.2 Subjective and objective simultaneously

In the trust model, the rule-bound deliberations represented by $b^* < b$ had an effect on objective fitness only through modifications of behaviour. To focus on the difference between $b$ and $b^*$ set $m = b^* - b < 0$ (recall also that we assumed that $b^* < 1 < b$).

In the preceding discussion, an individual who acted in exploitative ways in the second-mover role received the full objective payoff (although subjectively ranking it differently). In short the '$m$' leading to the subjective revaluation represented a purely subjective influence to which no objective payoff modification corresponded. However, since all other payoffs were subject to a dual interpretation, subjective and objective, why should not $m$ be Janus-faced as well?

Recall that the subjective payoff functions $u$, except for the parameter $b^*$, were assumed to be positive, linear transformations of the objective payoffs with a multiplicative factor $\lambda = 1$ and an additive $\beta = 0$. Now, for any $m$ we could conceivably allow that $m\lambda$ with $0 \leq \lambda \leq 1$ is the objective payoff corresponding to $m$. Then, unless $\lambda = 0$, the modified function is not anymore an element of the class of equivalent functional representations of the underlying objective payoffs.

---

[16] On the two-speed model of evolution see Güth and Kliemt 1998.

The parameter $\lambda \in [0,1]$ 'measures' how strict the separation between the objective and the subjective layer is. Less formally, the interpretation could in many cases capture the degree to which 'intrinsic' as opposed to 'extrinsic' motivation is operative as a cause of overt behaviour. With $\lambda = 0$, the objective payoff function remains unaltered, and intrinsic motivation must do all the work. With $\lambda = 1$, one might say that the motivational factor $m$ is backed up by extrinsic motivation in full.[17]

The preceding interpretation leads to a rather straightforward connection between the subjective and the objective sphere. It is superimposed on the interaction between the spheres described by the indirect evolutionary model.

As has been sketched so far the indirect evolutionary approach allows for an integration of otherwise separate theories in a single, though layered, conceptual scheme. The theoretical layers may be seen as different windows on 'one' world or be taken as representing a layered reality. Accounts of behaviour formulated in terms of 'rationalistic conscious maximization' and 'evolutionary interpretations' (see again Aumann 1998, p. 192) can be incorporated in the same conceptual scheme, and the relations between separate theoretical layers discussed.

### 6.5 ANALOGIES BETWEEN LAYERS OR INTERACTION OF LAYERS

It would be beyond the limits of our knowledge, and the scope of this chapter, to examine alternative positions with any reasonable claim to completeness. Therefore we will confine ourselves to a discussion of the view of Robert Aumann, who has written about similar issues, often with a different emphasis and conclusions.

#### *6.5.1 Robert Aumann speaking*

In fact, I find it somewhat surprising that our disciplines[18] have any relation at all to real behaviour. (I hope that most readers will agree that there is indeed such a relation, that we do gain *some* insight into the behaviour of *Homo sapiens* by studying *Homo rationalis*.) There is apparently some kind of generalized invisible hand at work. While in any given situation an individual may well act irrationally, there seems to be a cumulative effect of numbers and time and learning that

---

[17] Güth and Peleg (2001) consider both $\lambda = 0$ and $\lambda = 1$.
[18] Aumann is here referring to game and economic theory.

pushes people 'in general' in the direction of rational decision-making. It doesn't make people more and more rational, but as a given setting gets more and more common and familiar, it makes them act more and more rationally *in-that set-ting*. ... if one is careful not to expect too much, then *Homo rationalis* can serve as a model for certain aspects of the behaviour of *Homo sapiens*. (Aumann 2000, I, p. 12)

This passage is reminiscent of Schumpeter's statement quoted earlier. Unlike Schumpeter, Aumann can make use of a fully fledged game theory and concepts like Cournot-Nash equilibrium. Against the background of the pull versus push issue, this raises the question how the concepts of Nash equilibrium, of eductive decision theory, and the *evolutionarily stable strategies* (ESSs) of evolutionary theory are related to each other. Asking this does not amount to contesting the mathematical fact that ESS is a refinement of the Nash equilibrium concept. The issue is whether there is a framework in which the teleological and the non-teleological concepts can be integrated, and their role and relationships be better understood. And this is not a question about prediction and observation only; it is a dispute about a true understanding of how the world is constituted.

In this Aumann like many economists does not seem to agree with us. He says, that

game theorists argue too much about interpretations. Sure, your starting point is some interpretation of Nash equilibrium; but when one is doing science, you have a model and what you must say is: What do we *observe*? Do we observe Nash equilibrium? Or don't we observe it? Now, *why* we observe it, that's a prob-lem that is important for philosophers, but less so for scientists. (Aumann 1998)

### 6.5.2 *Speaking to Robert Aumann*

Aumann is right in that predictions play a crucial role in empirical sci-ence. However, he is wrong in treating predictive power as sufficient for accepting a theory. That the predictions of a theory are valid in an intended realm of application is a necessary but not a sufficient condition for sound empirical science. It is of the essence of science to inquire what the regularities and laws are that are responsible for the occurrence of explananda (observations).

Empirical scientists are interested in discriminating between merely potential and true explanations (i.e. those relying on genuine 'laws of nature'). We need to rely on predictions to find out whether or not theor-ies and hypotheses are valid. However, acknowledging the indispensable role of prediction does not mean that science is only about predicting future observations. Its objective is to predict and to explain on the basis of valid laws, i.e. the laws that are indeed responsible for observing the predicted effects.[19] In short, prediction and observation are tools of discrimination between theories and not the aim of scientific theory formation itself.[20]

With regard to the issue of optimization versus adaptation, it may be that adaptation leads to behaviour that is *as if* it were the outcome of the maximization of an objective function. If we observe that people behave as if maximizing, this is not a good reason to say that the theory predicting equilibrium behaviour as the result of teleological processes has been corroborated. Like Aumann, we believe that optimal results are often not brought about by conscious maximization. Maximizing is usually not the (proximate) cause, even when we observe equilibrium behaviour.

As far as explanation of phenomena is concerned, our theories do not lead us to an answer, but rather to a new question. As Aumann himself says:

In fact, the evolutionary interpretation should be considered the fundamental one. Rationality, Homo Sapiens, if you will, is a *product* of evolution. We have become used to thinking of the 'fundamental' interpretation as the cognitive rationalistic one, and of evolution as an 'as if' story, but that's probably less insightful.

Of course, the evolutionary interpretation is historically subsequent to the rationalistic one. But that doesn't mean that the rationalistic interpretation is the more basic. On the contrary, in science we usually progress, in time, from the superficial to the fundamental. The historically subsequent theory is quite likely to *encompass* the earlier ones.

The bottom line is that relationships like this are often trying to tell us something, and that we ignore them at our peril. (Aumann 1998, p. 192)

We agree with Aumann's conclusion that 'relationships like this are often trying to tell us something', but as our discussion shows we differ from him about what that may be. Where Aumann thinks in terms of 'acquisition' of one perspective by another more fundamental one, we tend to think in terms of a 'merger' of perspectives. We believe that the correct view is one in which, depending on the aim of the explanatory exercise,

---

[19] Knowledge of a true 'if-then' relation (possibly a probabilistic one) is also necessary if we intend to interfere in the course of the world. Then 'as if' is obviously not enough.

[20] Our somewhat orthodox realism would need further defence. This is beyond the scope of this chapter; see, however, Albert 1985; Mayo 1996.

both push and pull are recognized as causal factors and understood in their interaction.

### 6.6 CONCLUDING REMARKS

The preceding account of layered theory formation is a more or less philosophical exercise. Hard-nosed formal theorists, as well as empiricists, will perhaps tend to dismiss it as merely conceptual. But being conceptual does not render these ideas useless for the more practical concerns of social-theory formation. Fundamental conceptual orientations and some account of how things 'hang together' within social theorizing necessarily exert an influence on the day-to-day business of the social theorist.[21]

On the methodological plane, we do not agree with the implicit operationalism still in vogue among economists.[22] We believe that the chief task of science is to understand *why* we observe certain phenomena. Of course, this is not to deny that prediction of observable phenomena is crucial for science. Science would be devoid of empirical content if it could not predict certain observations and rule out others. Prediction is a means to separate true from false theories and explanations. But this does not mean that prediction is the chief aim of science nor that predictive content is an overriding criterion in assessing the quality of theories.

As far as empirical content is concerned, eductive models of maximization under constraints and evolutionary modelling are both somewhat weak. However, without them, and the *push* and *pull* that they describe, we will not be able to arrive at an adequate understanding of the world around us. The interaction of spheres must be taken into account as well. Indirect evolutionary modelling is *a* possible way to combine 'push and pull' systematically.

### REFERENCES

Ainslee, George 2002. *Break Down of the Will*, Princeton University Press.
Albert, Hans 1985. *Treatise on Critical Reason*, Princeton University Press.
Aristotle n.d. 'Physics, by Aristotle, Book II', *e-Books@Adelaide*, University of Adelaide website. http://etext.library.adelaide.edu.au/a/aristotle/a8ph/book2.html.
Aumann, Robert J. 1998. 'On the State of the Art in Game Theory: An Interview with Robert Aumann', *Games and Economic Behavior* 24: 181–210.

[21] Theorists guided by such conceptual orientations, besides Aumann himself, are for example, Binmore or Selten.
[22] Which of course echoes the orthodox methodological views of Friedman; see Friedman 1953/1966.

2000. *Collected Papers*, vols I and II, MIT Press, Cambridge, MA.

Berninghaus, Siegfried, Güth, Werner and Kliemt, Hartmut 2003. 'From Teleology to Evolution: Bridging the Gap between Rationality and Adaptation in Social Explanation', *Journal of Evolutionary Economics* 13(4): 385–410.

Binmore, Ken 1987. 'Modeling Rational Players', *Economics and Philosophy* 3: 179–214.

Brennan, Harold G. and Buchanan, James McGill 1985. *The Reason of Rules*, Cambridge University Press.

Brennan, Harold G., Güth, Werner and Kliemt, Hartmut 2003. 'Trust in the Shadow of the Courts', *Journal of Institutional and Theoretical Economics* 159(1): 16–36.

Buchanan, James M. and Tullock, Gordon 1962. *The Calculus of Consent*, University of Michigan Press, Ann Arbor.

Elster, Jon 1979. *Ulysses and the Sirens*, Cambridge University Press.

Frey, Bruno S. 1997. *Not Just For the Money: An Economic Theory of Personal Motivation*, Edward Elgar, Cheltenham.

Friedman, Milton 1953/1966. 'The Methodology of Positive Economics', in: Friedman (ed.), *Essays in Positive Economics*, Chicago University Press, pp. 3–46.

Güth, Werner, Kareev, Yaakov and Kliemt, Hartmut 2005. 'How to Play Randomly without Random Generator – The Case of Maximin Players', *Homo Oeconomicus* 22(2): 231–255.

Güth, Werner and Kliemt, Hartmut 1998. 'The Indirect Evolutionary Approach: Bridging the Gap between Rationality and Adaptation', *Rationality and Society* 10(3): 377–399.

2000. 'Evolutionarily Stable Co-operative Commitments', *Theory and Decision* 49: 197–221.

2007. 'The Rationality of Rational Fools', in: Peter and Schmid (eds.), *Rationality and Commitment*, Oxford University Press, pp. 24–149.

Güth, Werner, Kliemt, Hartmut and Peleg, Bezalel 2000. 'Co-evolution of Preferences and Information in Simple Games of Trust', *German Economic Review* 1(1): 83–110.

Güth, Werner and Ockenfels, Axel 2005. 'The Coevolution of Morality and Legal Institutions: An Indirect Evolutionary Approach'. *Journal of Institutional Economics* 1(2): 155–174.

Güth, Werner and Peleg, Bezalel 2001. 'When Will Payoff Maximization Survive? – An Indirect Evolutionary Analysis', *Journal of Evolutionary Economics* 11: 479–499.

Hardin, Russell 2007. *David Hume: Moral and Political Theorist*, Oxford University Press.

Lahno, Bernd 2002. *Der Begriff des Vertrauens*, Mentis, Paderborn.

Max Planck Institute of Economics n.d. 'Our Exercises of Indirect Evolution', Max Planck Institute of Economics website. www.econ.mpg.de/files/2011/Appendix_Exercises_on_indirect_evolution.pdf

Mayo, Deborah G. 1996. *Error and the Growth of Experimental Knowledge*, Chicago University Press.

Pearl, Judea 2000. *Causality: Models, Reasoning, and Inference*, Cambridge University Press.

Plott, Charles R. and Smith, Vernon L. (eds.) 2008. *Handbook of Experimental Economics Results*, North Holland, Amsterdam.

Samuelson, Larry 2001. 'Introduction to the Evolution of Preferences', *Journal of Economic Theory* 97: 225–230.

Schelling, Thomas C. 1984. *Choice and Consequence*, Harvard University Press, Cambridge, MA.

Schneider, Louis (ed.) 1967. *The Scottish Moralists on Human Nature and Society*, University of Chicago Press.

Schumpeter, Joseph A. 1959. *The Theory of Economic Development*, Harvard University Press, Cambridge, MA.

Spinoza, Baruch 1992. *Ethics*, Hackett, Indianapolis and Cambridge.

Stöckler, Manfred 1991. 'A Short History of Emergence and Reductionism', in: Agazzi (ed.), *The Problem of Reductionism in Science*, Kluwer, Dordrecht, pp. 71–90.

# CHAPTER 7

# Schelling formalized
## Strategic choices of non-rational behaviour
### David H. Wolpert and Julian Jamison

## 7.1 INTRODUCTION

### 7.1.1 Behavioural preference models

It is well-established that even in an anonymous single-shot game where players know they will never interact with their opponent(s) again, human players often exhibit 'non-rational' behaviour. (See Camerer 2003; Gächter and Herrmann 2009; and references therein.) To state this more precisely, often in an anonymous single-shot game where there are exogenously provided (often material) *underlying* preferences, humans do not maximize those underlying preferences. A lot of research has modelled such non-rational behaviour by hypothesizing that humans have *behavioural* preferences, which differ from their underlying preferences, and that they maximize behavioural preferences rather than underlying preferences. We will refer to such models as *behavioural preference models*. Different kinds of behavioural preference models arise for different choices of how to formalize the underlying and behavioural preferences. We will refer to the non-rational behaviour given by simultaneous maximization of every player's behavioural preferences as a *behavioural preference equilibrium* (BPE).

Perhaps the most prominent example of a behavioural preference model is the work on interdependent / other-regarding / social preferences (Bergstrom 1999; Kockesen et al. 2000; Sobel 2005). These models presume that people do not maximize expected underlying utility subject to play of their opponents, but instead maximize expected behavioural utility.

We would like to thank Raymond Chan, Nils Bertschinger, Nihat Ay, and Eckehard Olbrich for helpful discussion, as well as Lynn Conell-Price and the editors for valuable feedback on the manuscript.

Other work has explored behavioural preference models when the behavioural preferences are not expressible as expected utilities. An example is given by the (logit) quantal response equilibrium (QRE). Say we have a finite strategic form game where the mixed strategy of player $i$ over moves $x_i \in X_i$ is $q_i(x_i)$. Then under the QRE,

$$q_i(x_i) \propto e^{\lambda_i E(u_i|x_i)} \ \forall \text{ player } i. \tag{7.1}$$

The majority of work on the QRE since its introduction has treated it as a parametric model of non-rational behaviour. In this interpretation, $\lambda_i \in \mathbb{R}$ is a parameter describing 'how rational' or 'how smart' a player $i$ is, which is fit to experimental data (Camerer 2003). As $\lambda_i$ approaches infinity, player $i$ only puts weight on moves that are true best responses.

The QRE cannot be expressed as a behavioural preference model where behavioural preferences are expected values of commonly known utilities; in general the solution to Equation 7.1 is not a Nash equilibrium (NE) for any set of utilities. However, define the *free utility* of player $i$ by

$$\mathscr{F}(q_i) \equiv \lambda_i \mathbb{E}(u_i) + S(q_i), \tag{7.2}$$

where $S(q_i) \equiv -\sum_{x_i} q_i(x_i)\ln(q_i(x_i))$, is the Shannon entropy of $q_i$ (Cover and Thomas 1991; Mackay 2003). When all players $i$ simultaneously maximize their free utilities, we have a QRE (Wolpert 2009; Meginniss 1976). Accordingly, when players adopt a QRE with a particular set of rationalities $\{\lambda_i\}$, their mixed strategy profile is a BPE where all players simultaneously maximize their associated (behavioural) free utilities, rather than maximize the expected values of some behavioural utility functions (Meginniss 1976; Fudenberg and Kreps 1993; Anderson et al. 2001; Hopkins 2002; Fudenberg and Levine 1993; Shamma and Arslan 2004; Wolpert 2007).

Define an *objective function* as a map from the space of possible mixed strategy profiles to $\mathbb{R}$ (Wolpert 2009). As an example, given a utility function $u$, the expectation of $u$ under a mixed strategy profile is an objective function. Similarly, a free utility is an objective function. We will restrict attention to preferences that can be expressed as objective functions. So for us, a behavioural preference model explains human behaviour by hypothesizing that players maximize behavioural objective functions (subject to play of others) rather than maximizing underlying objective functions.

In this interdependent preferences and QRE experimental work, the task of the researcher is simply to ascertain the parameters of real-world behavioural objective functions from data. Two important issues are unaddressed in that work: how the players acquire common knowledge of one another's behavioural objective functions before start of play and why the parameters of the behavioural function have the values they do.

In this chapter we address this second issue. We start by noting that changing the values of parameters in the behavioural objective functions changes the equilibrium strategy profile. In particular, for a fixed set of behavioural objective function parameters for all players other than $i$, by varying the parameters of $i$'s behavioural objective function, we create a set of equilibrium profiles of the associated behavioural games. The profiles in that set can be ranked in terms of player $i$'s underlying objective function.

Our thesis is that humans learn over the course of their lifetimes which values of the parameters of their behavioural objective functions have the highest such rank in terms of their underlying objective function. In this way, the parameters of their behavioural objective function are determined endogenously, in a purely rational way, as the values that optimize their underlying objective functions.

### 7.1.2 Evolution of preferences

We begin by discussing earlier work that also addresses the issue of how players' behavioural objective functions are determined. This work is evolution of preferences (EOP), also known as 'indirect evolution' or 'two-timescale games' (Güth 1995; Huck and Oechssler 1999; Bester and Güth 2000; Samuelson 2001; Dekel et al. 2007; Heifetz et al. 2007).

Loosely speaking, EOP models human behaviour using two coupled strategic scenarios involving two different timescales. There is an exogenously provided set of players' pure strategy spaces and players have exogenously supplied underlying utility functions. Players choose a distribution across their space of pure strategies to play a 'short timescale' game. Critically, players do not choose mixed strategies to maximize their underlying utility function. (In this, they *appear* to be irrational.) Rather players choose their mixed strategies to maximize their behavioural utility functions.

In turn, the behavioural utility functions of the players are determined in a 'long timescale' process. The behavioural utility function of each player $i$ is set in this long timescale process so that resultant play in the short timescale game, according to players' *behavioural* utility functions, maximizes their *underlying* utility function. EOP concentrates on situations where the process occurring on the longer timescale is biological natural selection, operating on genomes that change across multiple generations (hence the name, 'evolution of preferences').

More precisely, EOP models all involve an infinite sequence of two-player games, $\gamma_t$. Each game has the same joint move space, $X$, a set of underlying player utilities over $X$, $\{u_i\}$, and an infinite population of players.

At iteration $t$ two players are randomly chosen from the infinite population to play the game $\gamma_t$. In contrast to much of evolutionary game theory, it is not the strategy $x_i \in X_i$ of each chosen player $i$ that is fixed at the beginning of iteration $t$, rather it is the *preferences* of player $i$ that are fixed. In other words, a behavioural utility function $b_i(t)$ from a set $B_i$ is fixed at the beginning of each iteration $t$ for both players chosen to play the game.

Given the joint behavioural utility $b(t) \equiv \{b_i(t)\}$, the game $\gamma_t$ proceeds in two stages. In the first stage, each player $i$ honestly signals their current behavioural utility $b_i(t)$ to the other player. In the second stage the players play $\gamma_t$ using the signalled joint behavioural utility $b(t)$. This means that in the second stage, the players adopt an NE of the *behavioural game* at $t$, i.e. of the game specified by the value of $b$ at $t$. So in EOP, the strategy of player $i$ in game iteration $t$ is not pre-fixed, but depends on signals from other players at $t$. Next, in analogy to much of evolutionary game theory, in EOP the mixed strategy profile over $X$ at iteration $t$ is evaluated under the actual, joint underlying utility $u$. This gives values of the underlying expected utilities of the players and the iteration $t$ ends.

In EOP, natural selection uses the expected underlying utilities from the final stage of the two-stage game at iteration $t$ as 'fitness' values, guiding the evolution of player behavioural utilities to the next iteration $t + 1$. This is analogous to how in much of evolutionary game theory the expected values of underlying utilities guide the evolution of player strategies. Accordingly, in EOP the appropriate equilibrium concept for the long timescale process in which the joint behaviour utility $b$ is determined is the evolutionarily stable strategy (ESS).

### 7.1.3 Difficulties with EOP

The central insight of EOP – that games occur on more than one timescale – is a powerful one. Unfortunately, calculations concerning EOP are quite complex. Thus much of the formal work in EOP restricts attention to a game (or set of coupled games) with a single symmetric utility function shared by all players. For similar reasons of tractability, much of EOP only concerns two-player games, often with only binary move spaces, and typically requires that the population be infinite. These simplifications restrict the ability of EOP to make predictions concerning experiments.

Other difficulties arise in the simple scenarios where one can do the calculations. For example, in some EOP games, the evolutionary dynamic process has no equilibrium at all and thus EOP cannot make predictions. A similar difficulty is that results in EOP typically vary with the initial characteristics of the population evolving.

These and other difficulties have prevented EOP from being used to analyse real-world behaviour. In particular, they have prevented EOP from being used to analyse either the results of fitting interdependent preference parameters to experimental data, or the results of fitting logit QRE parameters to experimental data.

### 7.1.4 Changing the long timescale in EOP

In this chapter we introduce a modification of EOP by allowing for broader sets of behavioural objective functions and changing the long timescale process to involve learning by a single individual across their lifetime, rather than natural selection occurring across multiple individuals' lifetimes. This modification allows us to replace the ESS solution concept with the NE.

With this modification, calculations become far more tractable, allowing us to extend analysis of two-timescale games to situations far beyond those that earlier EOP frameworks can analyse. Indeed, changing the long timescale allows us to make many theoretical predictions for the outcomes of game theory experiments, something EOP cannot do.

To avoid confusion with earlier EOP work where the long timescale process is natural selection, we use the term *persona* to indicate the behavioural objective function learned in the long timescale process we consider here. We refer to the associated two-timescale game as a *persona game*. Note that persona games assume that players can make binding commitments to adopt objective functions before the start of play that differ from their underlying objective functions, and that these commitments are made known to their opponents.[1]

---

[1] Many of the results go through even if these strict assumptions are relaxed somewhat. See Wolpert et al. (2010) for details. This assumption is the same one implicitly made in all the work on EOP, interdependent preferences and the QRE.

### 7.1.5 Chapter overview

We begin Section 7.2 by discussing how persona games appear to arise in the real world as binding 'moods' that people learn to adopt for real-world games. This discussion addresses the crucial issue of how players gain common knowledge of one another's binding behavioural objective functions at start of play, when they are told one another's underlying objective functions instead. We then provide a definition of persona games, along with illustrations of them.

In Section 7.3 we consider persona behavioural objective functions that are interdependent preference utility functions and show how persona games involving such personas can explain altruistic behaviour. We concentrate our analysis on the prisoner's dilemma (PD) and demonstrate how the Pareto efficient (cooperate, cooperate) outcome of the PD can arise as jointly rational behaviour of a persona game. Next we address a novel prediction concerning real-world play of the PD: an unavoidable trade-off between the utility gain to the players for jointly cooperating rather than jointly defecting and the robustness of a persona equilibrium that induces such joint cooperation. The sobering implication is that the greater the benefit of an equilibrium where people are altruistic, the more fragile that equilibrium. We end this section with some concrete predictions that can be tested in future experiments.

In Section 7.4 we show that empirical data on the traveller's dilemma (TD) (Basu 1994; Capra et al. 1999; Becker et al. 2005) is explained by persona games and our final comments are in Section 7.5.

## 7.2 PERSONA GAMES

### 7.2.1 Adopting and signalling personas

As illustrated by the QRE and interdependent preferences work, often people adopt and signal a behavioural objective function before the play of a game. Interestingly, it seems that this is often done so that the outcome of the resultant game is better for them than would otherwise be the case. This phenomenon of varying the persona we adopt to improve the outcome of the behavioural game is so widespread it has even entered common discourse, e.g. 'the workplace is full of chameleons who adopt a different persona each day' (Stern 2008).

In some circumstances a person may choose their persona, signal it to others, and commit to using it, all in a conscious manner. An example of this is whenever we interact with a very young child by acting dumber than we are. With adults, such conscious adoption and signalling of personas typically does not work, because the signalled persona usually is not believable to someone with fully developed social skills.

This does not mean that adults do not choose a persona, signal it to others, and commit to using it during play of a game. However, with adults it seems that all this is rarely done consciously, but rather in a semi-conscious or even unconscious manner. Such semi-conscious personas may be 'moods' which are induced by the person's knowledge of the game they are about to play. Such personas, once adopted, can only be discarded with substantial time and effort.

In most circumstances, which such persona is most suited to a given game is something that an individual learns over their entire lifetime, via personal experience, communication with others, etc., and that then gets manifested as a 'habit'.[2] All such learning processes occur on a long time-scale. In contrast, the effects of the learned persona transpire during the much shorter timescale on which the game is played.

In fact, there appear to be neurological feedback mechanisms forcing signalled and actual personas to match (Tom et al. 1991; Brinol and Petty 2003; Bechara and Damasio 2004; Tamir 2010). In general, if you act in a certain way, then you start to believe accordingly. This feedback helps stabilize personas, making them difficult to discard. (See the discussion in Frank (1987), Frith (2008), and of costly signalling in general in Spence (1973), Spence (1977) and Lachmann et al. (2001).) It would appear that humans have some ability, at least at a subconscious level, to adopt a persona that is binding on them and that they are forced to semi-honestly signal.

To give a graphic example, it appears that in the Ultimatum Game, the rate of rejection of low offers by responders when they have little time to decide whether to reject is far higher than when they have more time to decide (Grimm and Mengel 2010). In other words, on timescales so short that an adopted persona cannot be easily discarded (in this case the persona of refusing to allow oneself to be treated unfairly), such a persona governs an individual's publicly observable actions. However, on longer timescales, they can, with effort, discard the adopted persona, and instead maximize their utility.

---

[2] Indeed, this is the underlying thesis of the emerging literature in the psychology community on 'emotion regulation', which considers both biological, non-strategic reasons for which emotion a person should adopt, and the effect of those emotions on others (Tamir 2010).

The semi-conscious signal of such a persona can occur in tone of voice, body language, etc. Empirically, it is typically quite difficult for someone to consciously fake such a signal in a believable way. One indicator is the fact that in humans (unlike other species), the whites of the eyes are visible, so that where we look telegraphs our attention. Another is the anomalously large number of human facial muscles compared with those of other species; the 'purpose' of those muscles seems to be to signal emotions. (See, for example, the copious experimental data establishing unconscious control of facial expressions (Ekman 2007).)

Note though that the signal of a semi-consciously adopted persona does not necessarily have to be via body language. So long as a particular persona adopted by someone cannot be easily discarded, then that persona will be signalled via publicly observable actions by that person. From this perspective, body language signals of personas are simply adjuncts to persona signals in the form of publicly observable behaviour, helping reduce noise in those persona signals. The crucial thing is the ability to adopt a binding persona in the first place.

The question addressed in this chapter is what the strategic implications are of this ability to adopt a binding persona, assuming the person can find some way to signal that persona. Loosely speaking, how can a person exploit the fact that they can make binding commitments to adopt a persona? Is it ultimately advantageous to someone to adopt (and honestly signal) a finite rationality? Is it ultimately advantageous to someone to adopt (and honestly signal) an interdependent utility function? What kinds of 'non-rational' behaviour can be explained endogenously this way?

### 7.2.2  Definition of persona games

Variants of the persona-based explanation of non-rationality predate EOP, going back at least to the 1950s (Kissinger 1957; Schelling 1960; Raub and Voss 1990), and arguably to antiquity (Schelling 1960). We introduce a formal framework for this persona-based explanation of non-rationality, by modifying the long timescale of EOP. (See Wolpert et al. (2010) for all the technical details.)

In a persona game, as in EOP, each player $i$ has an associated underlying utility function $u_i : X \to \mathbb{R}$, where $X$ is the finite joint move space. Also as in EOP, each player $i$ has a set of possible behavioural objective functions, i.e. personas, which we write here as $A_i$. Each such persona maps the set of possible mixed strategy profiles over $X$ into $\mathbb{R}$. In addition to personas that are expected utilities, the usual choice in EOP, we also allow personas other than expected utilities.[3]

As in EOP, in persona games the underlying utility functions and sets of possible personas are used to define a two-stage game. In the first stage every player $i$ samples an associated distribution $P(a_i \in A_i)$ to get a persona $a_i$. Each player $i$ then honestly signals that $a_i$ to the other players. The mixed strategy profile of that NE determines the expected values of the players' underlying utility functions, also as in EOP.

EOP considers an infinite sequence of such two-stage games, where the personas are modified at the end of each game by natural selection. Here, persona games differ from EOP. In persona games we replace EOP's repeated games with a single game and we introduce the common knowledge assumption. In the scenarios we consider, this means that before signalling their personas, each player knows their own set $A_i$, the sets $\{A_j\}$ of the other players, and the underlying utility functions of all players. Each player $i$ uses this information to choose the distribution $P(a_i)$ to be sampled to generate the signalled $a_i$. They do so to maximize the associated expected value of their underlying utility, as evaluated under the NE of the behavioural games. This modification means that the distributions $P(a_i)$ themselves are NEs of the full, multistage game. In contrast, in EOP the distributions $P(a_i)$ change from one game to the next, only stabilizing if they settle on an ESS.

The players in the first stage of a persona game can be viewed as playing a game. Their joint move is the joint persona they adopt, $a$. The utility function of player $i$ in this game is the mapping from all possible values of $a$ to the expected underlying utility of the (NE of the) behavioural game specified by $a$. It is this full game that we have in mind when we use the term 'persona game'.

Note that in persona games, as in EOP, we assume that the underlying utility function $u_i$ of every player $i$ is provided exogenously, as is the set of $i$'s possible adopted personas, $A_i$. Physically, the determination of these exogenous factors may occur through an evolutionary process. However, as in the case of EOP, we do not model such processes here.

We refer to each $A_i$ as a *persona set*. The persona sets humans use in field experiments appear to be quite elaborate and it seems that they are often at least partially determined by social conventions and norms. It

---

[3] This extra breadth is needed, for example, to model the irrational player, who prefers a uniform mixed strategy to all other possible mixed strategies, rather than just being indifferent over their moves. See also von Widekind (2008).

also seems that people sometimes use a different persona set for different joint underlying utility functions $u$.

Here, for simplicity, we do not address such complexities. Nor do we address the issue of why some persona sets arise in human society but not others. (This limitation is shared with the work on interdependent preferences.) Rather we concentrate on some simple personas suggested by the interdependent preferences and QRE literatures, to investigate the explanatory power of persona games.

### 7.2.3 Illustrations of persona games

Adopting a persona that disagrees with one's true utility would seem to be non-rational. To illustrate how it can actually be rational, first consider a game involving two players, Row and Col, each of whom can choose one of two moves, (top, down) ({T,D}) for Row, and (left, right) ({L,R}) for Col. Let the utility function bimatrix $(u^R, u^C)$ be

$$\begin{bmatrix} (0,0) & (6,1) \\ (5,5) & (4,6) \end{bmatrix}, \tag{7.3}$$

where as usual, in each cell Row's utility comes first and Col's second. The joint (T,R) move is the only NE of this game. At that NE, $\mathbb{E}(u^C) = 1.0$. Now say that rather than being rational, Col adopted a persona where they were perfectly *irrational*. That is, they adopt a free utility with a $\lambda$ of 0. This means that they commit to choosing uniformly randomly between their two moves, with no evident concern for the resultant value of their utility function, and therefore no concern for what strategy Row adopts. Then Row would have expected utility of 3 for playing T, and of 4.5 for playing D. So if Row were rational, given that Col is irrational, Row would play D with probability of 1. Given that Col plays both columns with equal probability, this in turn would mean that $\mathbb{E}(u^C) = 5.5$.

So by being irrational rather than rational, Col has improved their expected utility from 1 to 5.5. Such irrationality by Col allows Row to play a move that Row otherwise wouldn't be able to play, and that ends up helping Col. This is true even though Col would increase expected utility by acting rationally rather than irrationally if Row's *strategy* were fixed (at D). The important point is that if Col were to act rationally rather than irrationally while Row's *rationality* were fixed (at full rationality), then Col would decrease expected utility. This phenomenon can be seen as a model of the common real-world scenario in which someone 'acts dumber than they are' (by not being fully rational), and benefits by doing so.

Now consider a game where each player has four possible moves rather than two, and the utility functions ($u^R$, $u^C$) are

$$\begin{bmatrix} (0,6) & (4,7) & (-1,5) & (4,4) \\ (-1,6) & (5,5) & (2,3) & (7,4) \\ (-2,1) & (3,2) & (0,0) & (5,-1) \\ (1,1) & (6,0) & (1,-2) & (6,-1) \end{bmatrix}. \tag{7.4}$$

An NE of this game is the joint pure strategy where Row plays the bottom-most move, and Col plays the left-most move. In other words, the 'joint rationality move' of (rational, rational) results in ultimate payoff of (1,1).

In contrast, say that both players are *anti-rational* in that they want to *minimize* their utility functions. Now an equilibrium occurs if Row plays the top-most row and Col plays the right-most column. The resultant ultimate payoff is (4,4) rather than (1,1).

There are also two intermediate cases, where one player is rational and one is anti-rational, (rational, anti-rational) and (anti-rational, rational). Equilibria for those two cases are given by the remaining two entries on the diagonal line through the bimatrix in Equation 7.4 extending from down-left to top-right. Plugging those values in results in the following bimatrix of what ultimate payoffs ensue for the four possible joint personas:

|  | Col rationality | |
|---|---|---|
|  | $-\infty$ | $+\infty$ |
| Row rationality | | |
| $-\infty$ | (4,4) | (3,2) |
| $+\infty$ | (2,3) | (1,1) |

$$\tag{7.5}$$

where full rationality is indicated by the value $+\infty$ and anti-rationality by $-\infty$.

This bimatrix defines a 'persona game', which controls what personas the players should adopt. In this persona game, $(-\infty, -\infty)$ is a dominant

NE. At that joint persona, neither player would benefit from changing to rational behaviour, no matter which persona their opponent adopted. Note in particular that $(+\infty, +\infty)$ is not an NE of the persona game. So if the players are sophisticated enough to play the persona game with each other rather than the underlying game, they will both act anti-rationally. Colloquially, in this game both players have an incentive to be 'self-destructive', regardless of whether their opponent is.

In this game, when both players are anti-rational, the resultant utility is better for both of them than when both are rational. So (anti-rational, anti-rational) is not only a jointly dominant choice of rationalities, it is also Pareto efficient. This is not always the case; there are games where (anti-rational, anti-rational) is the jointly dominant choice, but it is not efficient. The implication is that in some but not all games, good public policy would be to induce the players to try to be self-destructive.

### 7.2.4 Relation between persona games and EOP

Since persona game equilibria are NEs, all the powerful techniques for analyzing NEs can be used to predict what values of $P(a_i)$ arise in the real world. In contrast, in EOP the distributions analogous to $P(a_i)$ are equilibria of a dynamic evolutionary process, and often many techniques for analyzing NEs cannot be applied. This is why it is easier to generate theoretical results in the persona framework than in EOP. In particular, in the persona framework, one does not have to restrict any of the analysis to symmetric games involving few (e.g. two) players with small (e.g. binary) move spaces, as one often must in the EOP.

The use of NE techniques also means that every game in the persona framework has an equilibrium, so the persona framework can always make a prediction for how humans will behave, in contrast to EOP. Furthermore, being based on the NE, no initial characteristics of a population are relevant in the persona framework, and we make no physically impossible assumptions about such a population. Again, this contrasts with EOP.

As mentioned above, since behavioural utilities vary across the population in EOP, there must be some mechanism that before the start of the game at any instant $t$ forces each player $i$ to send an honest signal to all players $-i$ telling them the behavioural utility that player $i$ will use in that time-$t$ game. More precisely, recall that in the EOP $u_i$ is the underlying utility of player $i$ at time $t$, while $b_i(t)$ is the behavioural utility that $i$

will actually use to decide their move in the game played at $t$ (see Section 7.1.2). Define $s_i(t)$ as a signal that player $i$ sends (inadvertently or otherwise) to all players $-i$ before the start of the time-$t$ game. Then common knowledge for that time-$t$ game requires the existence of some mechanism that forces $s_i(t)$ to honestly reflect $b_i(t)$.

Precisely what this mechanism is that forces honest signalling of $b_i(t)$ is not specified in EOP. Nor is it specified how this mechanism comes to exist in the first place. However the mechanism operates, given that it enforces honest signalling of $b_i(t)$, there is no reason that player $i$ would not learn to take advantage of it. They would do this by strategically choosing what behavioural utility $b_i(t)$ to use – knowing that $b_i(t)$ will be honestly signalled to the other players – so as to optimize the resultant value of $b_i(t)$ in the time-$t$ game. In essence, persona games are an analysis of such strategic exploitation of the unavoidable honest signalling mechanism. In this sense, if EOP is a valid description of real behaviour, then so should be persona games. However, the persona game framework goes further. It can also be used to analyse scenarios where the signalling of personas is noisy or even non-existent (see Wolpert et al. 2010).

Finally, and perhaps most importantly, the persona framework offers explanations for apparent non-rationality even in anonymous single-shot games where the players know they will never again interact. This corroborates the conclusion in Henrich et al. (2002) that 'cooperation can (be explained), even among non-kin, in situations devoid of repeat interaction'. However, the persona framework shows that this conclusion holds even without punishment and genes for non-kin altruism (which have not been found on the human chromosome), whereas Henrich et al. (2006) assumes both of those. Cooperation can exist for purely self-interested reasons.

### 7.3 PERSONA GAMES WITH OTHER-REGARDING PERSONAS

To illustrate the breadth of persona games, we now consider personas for a player that involve the utilities of that player's opponents. Such personas allow us to model other-regarding preferences, e.g. altruism and equity biases. If a player benefits by adopting a persona with such an other-regarding preference in a particular game, then that other-regarding preference is actually optimal for purely *self*-regarding reasons.

### 7.3.1 Personas and altruism

Let $\{u_j : j = 1, ..., N\}$ be the utility functions of the original $N$-player underlying game. Have the persona set of player $i$ be specified by a set of distributions $\{\rho_i\}$, each distribution $\rho_i$ being an $N$-dimensional vector written as $(\rho_i^1, \rho_i^2, ..., \rho_i^N)$. By adopting persona $\rho_i$, player $i$ commits to playing the behavioural game with a utility function $\sum_j \rho_i^j u_j$ rather than $u_i$. So pure selfishness for player $i$ is the persona $\rho_i^j = \delta(i,j)$, which equals 1 if $i = j$, 0 otherwise. 'Altruism' then is a $\rho_i^j$ that places probability mass on more than one $j$.

To illustrate this, consider the two-player two-move underlying game with the following utility functions:

$$\begin{bmatrix} (2,0) & (1,1) \\ (3,2) & (0,3) \end{bmatrix}. \tag{7.6}$$

There is one joint pure strategy NE of this game, at (T,R). Say that both players $i$ in the associated persona game only have two possible pure strategies, $\rho_i^j \triangleq \delta(i,j)$ and $\rho_i^j \triangleq 1 - \delta(i,j)$, which we refer to as selfish ($\mathscr{E}$) and saint ($\mathscr{A}$), respectively. Under the $\mathscr{E}$ persona, a player acts purely in their own interests, while under the $\mathscr{A}$ persona, they act purely in their *opponent*'s interests.

For example, if Row chooses $\mathscr{E}$ while Col chooses $\mathscr{A}$, then the behavioural game equilibrium for the underlying game in Equation 7.6 is (D,L), since Row's payoff there is maximal. Note that this joint move gives both players a higher utility (3 and 2, respectively) than at (T,R), the behavioural game equilibrium when they both adopt the selfish persona. Continuing this way, we get the following pair of utility functions for the possible joint persona choices:

|  | Col $\rho$ | |
|---|:---:|:---:|
|  | $\mathscr{E}$ | $\mathscr{A}$ |
| Row $\rho$ | | |
| $\mathscr{E}$ | (1, 1) | (3, 2) |
| $\mathscr{A}$ | (0, 3) | (3, 2) |

(7.7)

The pure strategy NE of this persona game is ($\mathscr{E}$, $\mathscr{A}$), i.e. the optimal persona for Row to adopt is to be selfish, and for Col is to be a saint. Note that both players benefit by having Col be a saint. One implication is that Row would be willing to pay up to 2 to induce Col to be a saint. More surprisingly, Col would be willing to pay up to 1 to be a saint and work purely in Row's interests.

### 7.3.2 The prisoner's dilemma

In the case of the PD underlying game, other-regarding personas can lead the players in the behavioural game to cooperate. For example, say that each player $i$ can choose either the selfish persona or a 'charitable' persona, $\mathscr{L}$, under which $\rho_i$ is uniform (so that player $i$ has equal concern for their own utility and for their opponent's utility). Consider the PD where the utility function bimatrix $(u^R, u^C)$ is

$$\begin{bmatrix} (6,0) & (4,4) \\ (5,5) & (0,6) \end{bmatrix}, \tag{7.8}$$

so (defect, defect) is (R,T). Then the utility matrix for a charitable persona is

$$\begin{bmatrix} 3 & 4 \\ 5 & 3 \end{bmatrix}. \tag{7.9}$$

So, for example, if the row player is selfish and the column player is charitable, the behavioural game is

$$\begin{bmatrix} 6,3 & 4,4 \\ 5,5 & 0,3 \end{bmatrix}, \tag{7.10}$$

with an equilibrium at (defect, defect). The complete persona game is

|  | Player 2 persona | |
|---|---|---|
| Player 1 persona | $\mathscr{E}$ | $\mathscr{C}$ |
| $\mathscr{E}$ | (4, 4) | (4, 4) |
| $\mathscr{C}$ | (4, 4) | (5, 5) |

(7.11)

The efficient equilibrium of this persona game is for both players to be charitable, a choice that leads them to cooperate in the behavioural game. Note that they do this for purely self-centred reasons, in a game they play only once. This result might account for some of the experimental data showing a substantial probability for real-world humans to cooperate in such single-play games (Tversky 2004).

To investigate the breadth of this PD result, consider the fully general, symmetric PD underlying game, with utility functions

$$
\begin{bmatrix}
(\beta,\beta) & (0,\alpha) \\
(\alpha,0) & (\gamma,\gamma)
\end{bmatrix},
\tag{7.12}
$$

where (R,D) is (defect, defect), so $\alpha > \beta > \gamma > 0$. We are interested in what happens if the persona sets of both players are augmented beyond the triple {fully rational persona, the irrational persona, the anti-rational persona} that was investigated above, to also include the $\mathcal{L}$ persona. More precisely, we augment the persona set of both players $i$ to include a fourth persona $\rho_i u_i + (1 - \rho_i)u_{-i}$. For simplicity, we set $\rho_i$ to have the same value $s$ for both players.

Define

$$
R_1 \equiv \beta - s\alpha,
\tag{7.13}
$$

$$
R_2 \equiv \gamma - (1-s)\alpha,
\tag{7.14}
$$

$$
B \equiv \beta - \gamma.
\tag{7.15}
$$

Working through the algebra (see Wolpert et al. 2010), we first see that neither the non-rational nor the anti-rational persona will ever be chosen. We also see that for joint cooperation in the behavioural game (i.e. (L,T)) to be an NE under the $(\mathcal{L}, \mathcal{L})$ joint persona choice, we need $R_1 > 0$. If instead $R_1 < 0$, then under the $(\mathcal{L}, \mathcal{L})$ joint persona either player $i$ would prefer to defect given that $-i$ cooperates.

Note that $R_1$ can be viewed as the 'robustness' of having joint cooperation be the NE when both players are charitable. The larger $R_1$ is, then the larger the noise in utility values, confusion of the players about utility values, or some similar fluctuation would have to be to induce a pair of charitable players not to cooperate. Conversely, the lower $R_1$ is, the more 'fragile' the cooperation is, in the sense that the smaller a fluctuation would need to be for the players not to cooperate.

Given that $R_1 > 0$, we then need $R_2 > 0$ to ensure that each player prefers the charitable persona to the selfish persona whenever the other player is charitable. $R_2$ can also be viewed as a form of robustness, this time of the players both wanting to adopt the charitable persona in the first place.

Combining provides the following result:

Theorem 7.1 – Consider a two-player persona world $(X, U, A)$ where each $X_i$ is binary, $U$ is given by the generalized PD with payoff matrix in Equation 7.12, and each player $i$ has the persona set $A_i = \{U_i, sU_i + (1-s)U_j\}$ for some $0 \leq s \leq 1$. In the associated (unique, pure move) extended persona equilibrium, the joint persona move is $(\mathscr{L}, \mathscr{L})$ followed by (L,T) whenever $s \in \left(1 - \dfrac{\gamma}{\alpha}, \dfrac{\beta}{\alpha}\right)$.

For our range of admissible $s$ to be non-empty requires that $\gamma > \alpha - \beta$. Intuitively, this means that player $i$'s defecting in the underlying game provides a larger benefit to $i$ if player $-i$ also defects than it does if $-i$ cooperates. It is interesting to compare these bounds on $\alpha$, $\beta$ and $\gamma$ to analogous bounds, discussed in Nowak (2006), that determine when direct reciprocity, group selection, etc., can result in joint cooperation being an equilibrium of the infinitely repeated PD.

Now say that one changes the underlying game of Equation 7.12 by adding a penalty $-c < 0$ to the payoff of every player $i$ if they defect, giving the bimatrix

$$\begin{bmatrix} (\beta,\beta) & (0,\alpha - c) \\ (\alpha - c,0) & (\gamma - c,\gamma - c) \end{bmatrix}. \tag{7.16}$$

In other words, one introduces a material incentive $c$ to deter defection. Say that $cs > \gamma - (1-s)\alpha$, and that both $\gamma > c$ and $\alpha - \beta > c$. Then the new underlying game is still a PD, and the new $R_1$ is still positive, but the new $R_2$ is negative, where before it had been positive. So the persona equilibrium will now be (E, E). This establishes the following result:

Corollary 7.1 – Consider a two-player persona world $(X, U, A)$ where each $X_i$ is binary, $U$ is given by the generalized PD with payoff matrix in Equation 7.16, and each player $i$ has the persona set $A_i = \{U_i, sU_i + (1-s)U_j\}$ for some $0 \leq s \leq 1$. Then if $s \in \left(1 - \dfrac{\gamma}{\alpha}, \dfrac{\beta}{\alpha}\right)$, and $c = 0$, the extended persona equilibrium is $(\mathscr{L}, \mathscr{L})$ followed by (L,T). If instead $c$ is changed so that $cs > \gamma - (1-s)\alpha$, $\gamma > c$ and $\alpha - \beta > c$, then in the extended persona equilibrium both players defect.

Under the conditions of the corollary, for purely self-interested reasons, adding a material incentive that favours cooperation instead causes defection. This is true even though the players had cooperated before. We have an automatic 'crowding out' effect (Bowles 2008). This is the only formal explanation of crowding out in single-shot games as rational behaviour that we are aware of.

Return now to the $c = 0$ PD in Equation 7.12, so that if both players defect, each player's utility is $\gamma$. For this underlying game, when the extended persona game equilibrium is $(\mathcal{L}, \mathcal{L})$ followed by (L,T), the benefit to each player of playing the persona game rather than playing the underlying game directly is $B$. Comparing this to the formulas for $R_1$ and $R_2$ establishes the following:

> Corollary 7.2 – Consider a two-player persona world $(X, U, A)$ where each $X_i$ is binary, $U$ is given by the generalized PD with payoff matrix in Equation 7.12, and each player $i$ has the persona set $A_i = \{U_i, sU_i + (1 - s)U_j\}$ for some $0 \le s \le 1$. Then $R_1 + R_2 + B \le 1$.

This sobering result says that there are unavoidable trade-offs between the robustness of cooperation and the potential benefit of cooperation in the PD, whenever (as here) the underlying game matrix is symmetric and both players can either be selfish or charitable for the same value of $s$. The more a society benefits from cooperation, the more fragile that cooperation.

To understand this intuitively, note that having $R_2$ large means that both $\gamma$ and $s$ are (relatively) large. These conditions guarantee something concerning your opponent: they are not so inclined to cooperate that it benefits you to take advantage of them and be selfish. On the other hand, having $R_1$ large guarantees something concerning you: the benefit to you of defecting when your opponent cooperates is small.

### 7.3.3  Experimental predictions

There are many predictions our model of the PD makes that could be tested experimentally. One could test the predictions for what parameters of PD games do (not) result in cooperation. One could also test the predictions for what parameters do (not) result in crowding out. Note in both of these predictions the importance of non-anonymity, to allow signalling of personas. This results in an informal prediction, that in repeated game versions of the PD with anonymity, due to no persona signalling, NE play is more likely to arise quickly, and defection to arise.

## 7.4 COMPARISON TO EXPERIMENTAL DATA

To illustrate the persona framework, we provide an explanation for experimental data concerning TD (the traveller's dilemma) (Basu 1994; Capra et al. 1999; Goeree and Holt 1999; Rubinstein 2004; Becker et al. 2005; Basu 2007). TD models a situation where two travellers fly on the same airline with an identical antique in their baggage, and the airline accidentally destroys both antiques. The airline asks them separately how much the antique was worth, allowing them the answers {2,3, ..., 101}. To try to induce honesty in their claims, the airline tells the travellers that it will compensate both of them with the lower of their two claims, with a bonus of $R$ for the maker of the lower of the two claims, and a penalty of $R$ for the maker of the higher of the two claims.

To formalize TD, let $\Theta(z)$ be the Heaviside step function, $\Theta(z) = \{0,1/2,1\}$ for $z < 0$, $z = 0$ and $z > 0$, respectively. Then for both players $i$, the utility function in the TD underlying game is $u_i(x_i, x_{-i}) = (x_i + R)\Theta(x_{-i} - x_i) + (x_{-i} - R)\Theta(x_i - x_{-i})$, where $R$ is the reward/penalty (for making a low/high claim), $x_i$ is the monetary claim made by player $i$, and $x_{-i}$ is the monetary claim made by the other player. The NE of this game is (2,2), since whatever $i$'s opponent claims, it will benefit $i$ to undercut that claim by 1. However, in experiments (not to mention common sense) this NE never arises.

Even when game theorists play TD with one another for real stakes, they tend to make claims that are not much lower than 101, and almost never make claims of 2. When describing these results, Basu (2007) called for a formalization of 'the idea of behavior generated by rationally rejecting rational behavior ... to solve the paradoxes that plague game theory'.

Consider a persona game based on the $R = 2$ TD underlying game. It seems that real humans are sometimes irrational (i.e. purely random) and sometimes fully rational. So we choose those as the possible personas of the players, indicated by $\rho = 0$ and $\rho = \infty$, respectively. When both players are fully rational, the expected utility to both is the NE value, 2, i.e. $\mathbb{E}(u_i \mid \rho_1 = \infty, \rho_2 = \infty) = 2$, for both players $i$. Now say that player $i$ is rational while the other player is irrational. This results in the expected utility

$$\mathbb{E}(u_i \mid x_i\beta_i = \infty, \beta_{-i} = 0) = \frac{1}{100}\left(\left[\sum_{y=2}^{x_i-1}(y-2)\right] + x_i + \left[\sum_{y=x_i+1}^{101}(x_i+2)\right]\right)$$

for all of $i$'s possible underlying game moves $x_i$. The (integer) maximum of this is at $x_i \in \{97, 98\}$. The associated expected utility is $\mathbb{E}(u_i \mid \beta_i = \infty, \beta_{-i} = 0) \simeq 49.6$.

Since player $i$ is indifferent between those two moves in the behavioural game and player $-i$ plays a uniform mixed strategy regardless of $i$'s move, any behavioural game mixed strategy by $i$ over those two moves is an NE of the behavioural game. However, while the expected underlying utility to player $i$ is the same for either behavioural game move $x_i \in \{97,98\}$, the expected underlying utility to player $-i$ differs for those two moves. Accordingly, the expected underlying utility to player $-i$ will depend on the (arbitrary) mixing probability $s$ with which player $i$ chooses between $x_i \in \{97,98\}$ in the behavioural game. Similar calculations hold when it is Player 1 who is irrational and Player 2 who is rational. The mixing probability for player $-i$ which is analogous to $s$ will be written as $q$.

Combining, we can express the NE of the behavioural games corresponding to all four possible joint personas with the following set of matrices:

$$\begin{bmatrix} \left(\dfrac{6967}{200},\dfrac{6967}{200}\right) & \left(\dfrac{s2661+(1-s)2665}{50},\dfrac{2479}{50}\right) \\ \left(\dfrac{2479}{50},\dfrac{q2661+(1-q)2665}{50}\right) & (2,2) \end{bmatrix}, \quad (7.17)$$

where the ordering is (Player 1 objective, Player 2 objective), the top row is for $\beta_1 = 0$, the left column is for $\beta_2 = 0$, and both $q$ and $s$ range over the interval $[0.0,1.0]$.

We can now define the full persona game utility functions by uniformly averaging over $s$ and $q$. Those (rounded) values are

|  | Player 2 rationality | |
|---|---|---|
| Player 1 rationality | 0 | +∞ |
| 0 | (34.8,34.8) | (53.3,49.6) |
| +∞ | (49.6,53.3) | (2,2) |

(7.18)

This persona game has two pure strategy NEs, $(\rho_1, \rho_2) = (0,\infty)$ and $(\rho_1, \rho_2) = (\infty,0)$. The associated distribution $P(x_1)$ for the first of these rationality NEs is uniform. The associated $P(x_2)$ instead has half its mass on $x_2 = 97$, and half on $x_2 = 98$. The two distributions for the other pure strategy rationality NE are identical, just with $P(x_1)$ and $P(x_2)$ flipped.

There is also a symmetric mixed strategy NE of this persona game. To calculate the behavioural game distribution associated with that mixed rationality NE, write

$$P(x_1,x_2) = \sum_{\rho_1,\rho_2} P(x_1,x_2 \mid \rho_1,\rho_2)P(\rho_1,\rho_2)$$

$$= \sum_{\rho_1,\rho_2} P(x_1,x_2 \mid \rho_1,\rho_2)P(\rho_1)P(\rho_2),$$

(7.19)

where each of the four distributions $P(x_1, x_2|\rho_1, \rho_2)$ (one for each $(\rho_1, \rho_2)$ pair) is evaluated by uniformly averaging the multiple behavioural game NE for that $(\rho_1, \rho_2)$ pair. For both players $i$ the associated marginal distribution is given by $P(x_i) = \sum_{\rho_1,\rho_2} P(x_i \mid \rho_1,\rho_2)P(\rho_1)P(\rho_2)$. Plugging in gives the result that both rationality players choose $\rho = 0$ with probability .78. The associated marginal distributions $P(x_i)$ are identical for both $i$'s: $P(x_i = 2) \simeq 5.8\%$, $P(x_i = 97) = P(x_i = 98) \simeq 9.5\%$, and $P(x_i) \simeq 0.8\%$ for all other values of $x_i$. (Note that because $P(\rho_1, \rho_2)$ is not a delta function, $P(x_1,x_2) \neq P(x_1)P(x_2)$.)

At such a mixed strategy NE of the persona game, the players randomly choose among some of their possible personas. Formally, the possibility of such a mixed NE is why persona games always have equilibria, in contrast to EOP. Empirically, such an NE can be viewed as a model of 'capricious' behaviour by humans.

In general, to compare experimental data with the predictions of game theory when there are multiple equilibria is an informal exercise. This is just as true here, with the multiplicity of persona game equilibria. We simply note that if we uniformly average over the three NE of the persona game, we get a $P(x)$ that is highly biased to large values of $x$. This agrees with the experimental data recounted above.

## 7.5 FINAL COMMENTS

Humans sometimes exhibit what appears to be non-rational behaviour when they play non-cooperative games with others (Camerer 2003; Kahneman 2003; Camerer and Fehr 2006). One response to this fact is simply to assert that humans are non-rational, and leave it at that. Under this response, essentially the best we can do is catalogue the various types of non-rationality that arise in experiments. Inherent in this response is the idea that 'science stops at the neck', that somehow logic suffices to explain the functioning of the pancreas but not of the brain.

There has been a lot of work that implicitly disputes this, and tries to explain the apparent non-rationality of humans as actually being rational, if the strategic problem is appropriately reformulated. The implicit notion in this work is that the apparent non-rationality of humans in experiments does not reflect 'inadequacies' of the human subjects. Rather it reflects inadequacies in us scientists, in our presumption to know precisely what strategic scenario the human subjects are considering when they act. From this point of view, our work as scientists should be to try to determine just what strategic scenario *really* faces the human subjects, as opposed to the one that *apparently* faces them.

One body of work that adopts this point of view is evolutionary game theory. The idea in evolutionary game theory is that humans (or other animals) really choose their actions in any single instance of a game to optimize results over an infinite set of repetitions of that game, not to optimize it in the single instance at hand. The persona framework is based on the same point of view that the apparent game and the real game differ.

Although persona games as defined here require agents to be able to both commit to a persona and then publicly and credibly signal that persona to others, these requirements can be relaxed. In particular, imperfect (noisy) signalling induces the same types of incentives over the choice of personas, since there is no direct payoff liability in the initial persona-choice stage of the game. In fact, personas are even relevant in a world in which no signalling is possible, as long as there is some cost to changing personas – in that case observation of previous actions serves the same role as imperfect signalling.

Persona games provide a very simple justification for irrationality with very broad potential applicability. They also make quantitative predictions that can often be compared with experimental data. While here we have only considered personas involving degrees of rationality and degrees of altruism, there is no reason not to expect other kinds of persona sets in the real world. Risk aversion, uncertainty aversion, reflection points, framing effects, and all the other 'irrational' aspects of human behaviour can often be formulated as personas.

Even so, persona games should not be viewed as a candidate explanation of all non-rational behaviour. Rather they are complementary to other explanations, for example those involving sequences of games (like EOP).

# REFERENCES

Anderson, S. P., J. K. Goeree and C. A. Holt 2001. 'Minimum-effect coordination games: stochastic potential and logit equilibrium', *Games and Economic Behavior* 34: 177–199.

Basu, K. 1994. 'The traveler's dilemma: paradoxes of rationality in game theory', *American Economic Review* 84: 391–5.

2007. 'The traveler's dilemma', *Scientific American* 296(6): 90–5.

Bechara, A. and A. Damasio 2004. 'The somatic marker hypothesis: a neural theory of economic decision', *Games and Economic Behavior* 52: 336–72.

Becker, T., M. Carter and J. Naeve 2005. 'Experts playing the traveler's dilemma', Universitat Hohenheim Nr. 252/2005.

Bergstrom, T. 1999. 'Systems of benevolent utility functions', *Journal of Public Economic Theory* 1: 71–100.

Bester, H. and W. Güth 2000. 'Is altruism evolutionarily stable?' *Journal of Economic Behavior and Organization* 34: 193–209.

Bowles, S. 2008. 'Policies designed for self-interested citizens may undermine "the moral sentiments"', *Science* 320: 1605.

Brinol, P. and R. Petty 2003. 'Overt head movements and persuasion: a self-validation analysis', *Journal of Personality and Social Psychology* 84: 1223–39.

Camerer, C. F. 2003. *Behavioral Game Theory: Experiments in Strategic Interaction*. Princeton University Press.

Camerer, C. F. and E. Fehr 2006. 'When does economic man dominate social behavior?', *Science* 311: 47–52.

Capra, C. M., J. K. Goeree, R. Gomez and C. A. Holt 1999. 'Anomalous behavior in a traveler's dilemma game', *American Economic Review* 19: 678–90.

Cover, T. and J. Thomas 1991. *Elements of Information Theory*. Wiley-Interscience, New York.

Dekel, E., J. Ely, and O. Yilankaya 2007. 'Evolution of preferences', *Review of Economic Studies* 74: 685–704.

Ekman, P. 2007. *Emotions Revealed*. Holt Paperbacks.

Frank, R. H. 1987. 'If homo economicus could choose his own utility function, would he want one with a conscience?' *American Economic Review* 77(4): 593–604.

Frith, C. 2008. 'Social cognition', *Philosophical Transactions of the Royal Society B*, doi:10.1098/rstb.2008.0005.

Fudenberg, D. and D. Kreps 1993. 'Learning mixed equilibria', *Games and Economic Behavior* 5: 320–67.

Fudenberg, D. and D. K. Levine 1993. 'Steady state learning and Nash equilibrium', *Econometrica* 61(3): 547–73.

Gächter, S. and B. Herrmann 2009. 'Reciprocity, culture and human cooperation: previous insights and a new cross-cultural experiment', *Philosophical Transactions of the Royal Society B* 354: 791–806.

Goeree, J. K. and C. A. Holt 1999. 'Stochastic game theory: for playing games, not just doing theory', *Proceedings National Academy of Sciences* 96: 10564–67.

Grimm, V. and F. Mengel 2010. 'Let me sleep on it: delay reduces rejection rates in ultimatum games', Maastricht University library website. http://edocs. ub.unimaas.nl/loader/file.asp?id=1491.

Güth, W. 1995. 'An evolutionary approach to explaining cooperative behavior by reciprocal incentives', *International Journal of Game Theory* 24: 323–44.

Heifetz, A., C. Shannon and Y. Spiegel 2007. 'The dynamic evolution of preferences', *Economic Theory* 32: 251–86.

Henrich, J., R. McElreath, A. Barr, J. Ensminger, C. Barrett, A. Bolvanatz, J. C. Cardenas, M. Gurven, E. Gwako, N. Henrich, C. Lesorogol, F. Marlowe, D. Tracer and J. Ziker 2002. 'Costly punishment across human societies', *Science* 312: 1767–70.

Hopkins, Edward 2002. 'Two competing models of how people learn in games', *Econometrica* 70: 2141–66.

Huck, S. and J. Oechssler 1999. 'The indirect evolutionary approach to explaining fair allocations', *Games and Economic Behavior* 28: 13–24.

Kahneman, D. 2003. 'Maps of bounded rationality: psychology of behavioral economics', *American Economic Review* 93: 1449–75.

Kissinger, H. 1957. *Nuclear Weapons and Foreign Policy*, Harper & Brothers.

Kockesen, L., E. Ok and R. Sethi 2000. 'The strategic advantage of negatively interdependent preferences', *Journal of Economic Theory* 92: 274–99.

Lachmann, M., C. Bergstrom and S. Szamado 2001. 'Cost and conflict in animal signals and human language', *Proceedings of the National Academy of Sciences* 98: 13189–94.

MacKay, D. J. C. 2003. *Information Theory, Inference, and Learning Algorithms*. Cambridge University Press.

Meginniss, J. R. 1976. 'A new class of symmetric utility rules for gambles, subjective marginal probability functions, and a generalized Bayes' rule', *Proceedings of the American Statistical Association, Business and Economics Statistics Section*, pp. 471–6.

Nowak, M. A. 2006. 'Five rules for the evolution of cooperation', *Science* 314: 1560–3.

Raub, W. and T. Voss 1990. 'Individual interests and moral institutions', in M. Hechter, K.-D. Opp and R. Wippler (eds.) *Social Institutions, Their Emergence, Maintenance and Effects*. Walter de Gruyter, Berlin, pp. 81–118.

Rubinstein, Ariel 2004. 'Instinctive and cognitive reasoning: a study of response times', Tel Aviv University website. http://arielrubinstein.tau.ac.il/papers/ Response.pdf.

Samuelson, L. (ed.) 2001. *Evolution of Preferences*, special issue, *Journal of Economic Theory* 97.

Schelling, T. C. 1960. *The Strategy of Conflict*. Harvard University Press, Cambridge, MA.

Shamma, J. S. and G. Arslan 2004. 'Dynamic fictitious play, dynamic gradient play, and distributed convergence to Nash equilibria', *IEEE Transactions on Automatic Control* 50(3): 312–27.

Sobel, J. 2005. 'Interdependent preferences and reciprocity', *Journal of Economic Literature* 43: 392–436.

Spence, M. 1973. 'Job market signaling', *Quarterly Journal of Economics* 87: 355–74.

1977. 'Consumer misperceptions, product failure and product liability', *Review of Economic Studies* 44: 561–72.

Stern, S. 2008. 'Be yourself but know who you are meant to be', *Financial Times*, 17 March.

Tamir, M. 2010. 'The maturing field of emotion regulation', *Emotion Review* 3(1): 3–7.

Tom, G., P. Pettersen, T. Lau, T. Burton and J. Cook 1991. 'The role of overt head movement in the formation of affect', *Basic and Applied Social Psychology* 12: 281–9.

Tversky, Amos 2004. *Preference, Belief, and Similarity: Selected Writings*. MIT Press, Cambridge, MA.

von Widekind, S. 2008. *Evolution of Non-expected Utility Preferences*. Springer, New York.

Wolpert, D. H. 2007. 'Predicting the outcome of a game', Submitted. See arXiv. org/abs/nlin.AO/0512015.

2009. 'Trembling hand perfection for mixed quantal / Best Response equilibria', *International Journal of Game Theory* 8(4): 539–551.

Wolpert, D. H., J. Jamison, D. Newth and M. Harre 2010. 'Schelling formalized: Strategic choices of non-rational personas', Working Paper 10-10, Federal Reserve Bank of Boston website. www.bostonfed.org/economic/wp/wp2010/wp1010.pdf.

CHAPTER 8

# Human cooperation and reciprocity

## Jack Vromen

### 8.1 INTRODUCTION

Proponents of strong reciprocity (henceforth SR) often emphasize that whereas much seemingly altruistic behavior is ultimately selfish on standard evolutionary explanations of altruism, SR is genuinely altruistic. Proponents of SR furthermore hold that it is adaptive behavior. They argue that SR could not possibly have evolved via standard individualist pathways, such as compensating future fitness benefits, as in reciprocal altruism (RA), or fitness benefits accruing to relatives (kin selection), and that it therefore must have evolved via some other pathway, such as group selection. Some opponents of SR agree that SR is genuinely altruistic, but argue that it is maladaptive behavior (Burnham and Johnson 2005). Other opponents of SR agree that it might be adaptive behavior, but argue that if it is adaptive behavior it is not genuinely altruistic (West et al. 2007).

A substantive part of this chapter is devoted to a critical scrutiny of this debate. I argue that it is absolutely essential to keep the notions of evolutionary and psychological altruism apart. Although both camps in the debate acknowledge this explicitly, I point out that the notions are sometimes conflated.[1] Once we clearly distinguish between evolutionary

Thanks go to participants of the Evolution, Co-operation, and Rationality conference in Bristol (18–20 September 2009) and to Herbert Gintis and especially Samir Okasha for useful comments on earlier drafts.

[1] This conflation is by no means confined to proponents of SR only. Opponents of SR also sometimes fail to keep the two apart. Consider for example Binmore's ambiguous statement: "Why do they feel a sense of duty? Ultimately, because they would be censored by their community if they were to fail to carry out the role assigned to them by the current social contract" (Binmore 2004, p. 29). What is ambiguous here is that Binmore might be referring to an ultimate explanation of why a sense of duty evolved in the first place (i.e., people without a sense of duty suffered fitness disadvantages). But he might also be referring to a proximate explanation of why people feel a sense of duty: because they fathom that a sense of duty is instrumental to avoiding being censored by their community.

and psychological altruism, the issue of whether SR is ultimately genuinely altruistic can be unpacked into two separate issues that in principle can be resolved clearly and independently of each other. I argue that the difference between SR and RA is much less clear-cut than SR's opponents suggest. It is not clear that SR is evolutionarily and psychologically altruistic. This inconclusiveness need not undermine the case for SR, however. I argue that it is unwise for proponents of SR to try to differentiate it from standard explanations of human cooperation in terms of altruism and selfishness. It raises unnecessary suspicion and irritation in onlookers who are in principle sympathetic to the idea that SR covers a significant part of human cooperation. And it diverts attention from more important issues. At the end of the chapter I identify a few issues that I think are more important for SR's cause than the issue of whether or not it is ultimately selfish.

## 8.2 WHAT IS SR?

I take the first introductory chapter in Gintis et al. (2005) to be exemplary for the way in which proponents of SR contrast it with RA. Gintis et al. start out by discussing what they call the model of the self-regarding actor that long dominated economics. Although they do not give a precise definition, it is clear that they have a model in mind in which actors are assumed to be concerned solely with serving their own interest, where the latter means increasing or maximizing their own material welfare. They then argue that a similar situation has existed in human biology. For decades biologists believed that two theories, the theory of inclusive fitness and of RA, were sufficient to explain human cooperation. "Trivers followed Hamilton in showing that even a selfish individual will come to the aid of an unrelated other, provided that there is a sufficiently high probability that the aid will be repaid in the future. ... These theories convinced a generation of researchers that, except for sacrifice on behalf of kin, what appears to be altruism (personal sacrifice on behalf of others) is really just long-run material self-interest" (ibid., p. 7).

But there is now massive experimental evidence, Gintis et al. go on to argue, supporting the ubiquity of non-self-regarding motives. One salient example of this is provided by SR: "Strong reciprocity is *a predisposition to cooperate with others, and to punish (at personal cost, if necessary) those who violate the norms of cooperation, even when it is implausible to expect that these costs will be recovered at a later date*" (ibid., p. 8, italics in the original). Thus, SR consists of two parts, we could say: strong positive reciprocity (also sometimes called altruistic rewarding), cooperating with others who likewise comply with the norms of cooperation; and strong negative reciprocity (also called altruistic punishment), punishing others who violate the norms of cooperation. What is striking about SR is that it manifests itself also in anonymous nonrepeated situations. This is puzzling from the perspective of evolutionary theory, since evolutionary theory suggests that this kind of self-sacrificing behavior (for which there is no "expectation" that the costs will be repaid later) should die out. Gintis et al. write: "... in a population of individuals who sacrifice for others, if a mutant arises that does not so sacrifice, that mutant will spread to fixation at the expense of its altruistic counterpart. Any model that suggests otherwise must involve selection on a level above that of the individual" (ibid., p. 8). Models of group selection are needed to explain how SR could have evolved.

Gintis et al. (2005) thus start by arguing that RA assumes that individuals are selfish in the sense that they are led by self-regarding motives. Reciprocal altruists only help others, it is argued, if they expect net benefits in due time from doing so. By contrast, SR is argued to assume altruistic individuals with other-regarding motives. Strong reciprocators are predisposed to help others also if this is costly for them and if they cannot expect these costs to be recovered in the future. As Gintis et al. describe the alleged selfishness of RA in terms of a particular sort of motive, it seems they have *psychological selfishness* in mind. As a first and rough approximation, psychological selfishness is the doctrine that the ultimate desires or motives that individuals act upon are self-regarding. But when Gintis et al. move on and argue that unlike RA, SR is puzzling from the perspective of evolutionary theory, it seems they have *evolutionary selfishness* in mind. Evolutionary altruism is defined not in terms of expected results, but in terms of actual fitness consequences. A behavior is called evolutionarily altruistic if it decreases the actor's fitness and increases the recipient's fitness. RA does not pose a puzzle for evolutionary theory, Gintis et al. argue, because it is not really evolutionarily altruistic: initial fitness costs incurred by the actor by rendering a favor to cooperators are repaid later by reciprocating acts by the recipients. By contrast, strong reciprocators are evolutionarily altruistic. Strong reciprocators really do sacrifice for others without initial fitness costs being recouped later.

In short, it seems that Gintis et al. are arguing that RA is both psychologically and evolutionarily selfish. And since they contrast SR with RA on both counts, it seems that they present SR as both psychologically and evolutionarily altruistic.

## 8.3 IS RA REALLY EVOLUTIONARILY AND PSYCHOLOGICALLY SELFISH?

It seems that the main reason for Gintis et al.'s arguing that RA is evolutionarily selfish is that RA does not really decrease the actor's individual (or personal) fitness. Initially the actor does incur an individual fitness cost by cooperating (rather than noncooperating) with others. If the actor had not cooperated, the actor's individual fitness would have been higher. But this decrease in individual fitness is only temporary. The momentary decrease in individual fitness is compensated for later by reciprocating acts of the recipients. Thus the lifetime individual fitness of the actor is not decreased but increased. This, we could say, is the crux of the ultimate explanation of momentary fitness-decreasing behavior that RA provides: the behavior "ultimately" yields individual fitness advantages for the actor.

This line of argument is defensible. If we insist that some behavior is evolutionarily altruistic only if it decreases the actor's individual fitness not only temporarily, but over the actor's lifetime, then RA comes out as evolutionarily selfish. This is also the line taken in West et al. (2007). West et al. propose to sharpen the standard definition of "evolutionary altruism" as follows: behavior that decreases the actor's *lifetime fitness* and increases the recipient's fitness. In West et al.'s helpful terminology, RA explains behavior that reduces the actor's fitness and increases the recipient's fitness by referring to the *direct* compensating fitness effects for the actor: fitness benefits that accrue to the actor later in the actor's life.

So far so good. But is RA also psychologically selfish? Gintis et al. (2005) clearly think so. Gintis et al. argue that unlike reciprocal altruists, strong reciprocators engage in costly cooperative and punishing behavior even when they cannot expect the costs to be recovered at a later date. The intended contrast here is that reciprocal altruists are willing to engage in costly cooperative and punishing behavior only if they can expect the costs to be recovered later (Fehr and Henrich 2003, pp. 57–8 and 61; Gintis et al. 2005, p. 6). Reciprocal altruists do not engage in costly cooperative and punishing behavior in anonymous one-shot games, because they believe the net long-run benefits for them are negative. They do not extend their costly behaviors to such situations, because they have self-regarding motives and understand that in such situations there is nothing to be gained. By contrast, strong reciprocators also know that they cannot expect to reap future benefits from engaging in such behavior, but they engage in it nonetheless. For Gintis et al. this testifies to the existence of other-regarding ultimate motives in strong reciprocators.

But in fact RA, properly understood as an ultimate explanation of conditional cooperation, does not imply that the actors involved are ultimately led by self-regarding motives. Actors engaged in costly helping behavior might be driven by a variety of other motives. They might like cooperating with other cooperators, for example. Or they may be led by a sense of fairness, prescribing that cooperation by others ought to be answered by cooperation by oneself. As long as the fitness advantages foregone now by cooperating with other cooperators are sufficiently paid back in the future by the recipients, conditional cooperation could have evolved. The cooperation at issue is conditional on what others did before and not necessarily on what others are expected to do in the future. The type of conditionality involved does not single out one unique proximate cause.

Consider Axelrod and Hamilton's 'tit for tat' (TfT), for example, which is often seen as a version of RA. It is tempting to understand TfT as involving a forward-looking calculating attitude: by answering your cooperation in the last round with my cooperation I hope to encourage you to continue to cooperate with me and thereby to benefit myself from this. And by answering your defection in the last round with my defection I hope to discourage your continuing to defect so as to benefit myself from this. But in fact such forward-looking calculation is neither implied nor required for TfT to be displayed. Strictly speaking, TfT involves no more than "start to cooperate and subsequently do what the other individual did in your last encounter." As for example Gigerenzer (2007) argues, it is enough for TfT to work that the following three building blocks are in place: cooperate first; keep a memory of size one; and imitate your partner's last behavior.[2]

Gintis et al. (2005) do not show that reciprocal altruists are ultimately driven by self-regarding motives. They rather take this for granted. One reason for this might be that they mix up ultimate and proximate explanation. Properly understood as an ultimate explanation, RA shows that cooperation with nonkin could have evolved because, and only insofar as, it does not decrease the actor's lifetime fitness. If we understand RA as a proximate explanation, however, this seems to translate naturally into the thesis that reciprocal altruists engage in cooperation only if they

---

[2] This is not to say that TfT is cognitively undemanding. It seems TfT is observed only in a few, cognitively sophisticated species (Hammerstein 2003; Stevens and Hauser 2004).

Figure 8.1. Superior fitness as eventual outcome: the causal chain of functionality.

expect that they serve their own lifetime fitness by doing so. Everybody agrees that ultimate explanations and proximate explanations should not be confused (Mayr 1961; Tinbergen 1963). But especially with respect to the debate over evolutionary altruism, the two are easily confused nonetheless. The reason for this, I submit, is that the sense of "ultimate" in ultimate explanations of evolutionary altruism tends to be conflated with the sense of "ultimate" in debates over psychological selfishness and psychological altruism.

"Ultimate" in ultimate explanations of evolutionary altruism refers to fitness benefits that must materialize to make it possible for momentary fitness-decreasing behavior to have evolved. Eventually, "at the end of the day," compensating fitness benefits must have fallen to actors (or their kin). Otherwise, momentary fitness-decreasing behavior could not have evolved.[3] The ultimate explanation provided in RA is that later fitness benefits for the actor generated by reciprocating acts by recipients make up for momentary fitness costs incurred by the actor. This is depicted in Figure 8.1. The arrow in the picture stands for a causal relation. As this is all about actual consequences of behavior, not about motives or reasons, let us call this the chain of *functionality*.

Especially in debates over psychological altruism it is tempting to mistake the chain of functionality for one of *instrumentality*. What should eventually be attained for some behavior to have evolved, superior lifetime fitness (or something of the sort), is then interpreted as the ultimate motive or desire that the actor is led by in his behavior. Since the desire to maximize one's own lifetime fitness is clearly a self-regarding desire, the behavior would appear to be an instance of psychological selfishness. Helping others is then seen as instrumental for satisfying the ultimate desire. This means in effect that the arrow in Figure 8.1 is reinterpreted

---

[3] Proponents of SR might object that this falsely takes for granted that there is no selection on a level above that of the individual. This issue will be further discussed below, in the section "Is SR psychologically altruistic?"

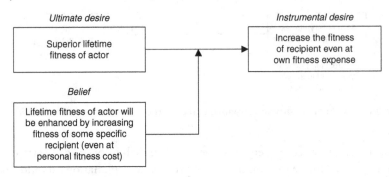

Figure 8.2. Superior fitness as ultimate desire: an inferential reasoning scheme.

as an inference made by the actor in a process of practical reasoning. Actually, if we want to depict such a psychological process as a causal one of practical reasoning, we should put things in reverse order. On the upper left side we would have the ultimate desire to achieve superior lifetime fitness. Via a causal process of practical reasoning, the actor would, if he in addition holds a suitable belief, infer from this ultimate desire an instrumental desire to increase the fitness of some recipient at his own fitness expense. Coupled with the right sort of belief, finally, this instrumental desire would issue, through another process of practical reasoning, in the fitness-sacrificing act. We would have something like the arrangement in Figure 8.2.

Though tempting, understanding the causal chain of functionality as an inferential scheme in a practical reasoning process should be resisted. The causal chain of functionality depicted in Figure 8.1 can materialize without the inferential scheme depicted in Figure 8.2 being in place. Attaining superior lifetime fitness need not be the *summum bonum* for actors for the causal chain of functionality to materialize. Actors might instead be led by a sense of justice, for example, prescribing that cooperative behavior by others ought to be rewarded and that noncooperative behavior ought to be punished. Furthermore, it is not necessary that the psychological process that leads actors to their conditional rewarding and punishing behavior be a practical reasoning process. As Trivers (1971) suggests, actors involved in RA might be led more directly by emotions such as generosity and gratitude than by a more elaborate and sophisticated practical reasoning process. Indeed, at times Trivers seems to suggest that it is not likely that elaborate processes of practical reasoning underlie RA. Altruistic acts performed on the basis of a calculating, self-serving motive would be less reliable and be distrusted more readily by others than altruistic acts based on emotions such as generosity and gratitude (Trivers 1971, pp. 50–51).

Trivers clearly did not conflate ultimate and proximate explanation. He neatly separates the issue of the general conditions that have to be met for RA to evolve and the discussion (in a separate section at the end of his paper) of what he calls "[t]he psychological system underlying human reciprocal altruism." Of special interest is Trivers' treatment of *moralistic aggression*. Trivers argues that it is likely that punitive emotions such as anger and indignation evolved to protect altruistic actors from being exploited by nonreciprocating individuals. Indeed, anthropological evidence indicates that a great deal of aggression revolves around real or imagined injustices (and cases of unfairness and lack of reciprocity). Often the severity of aggression seems out of all proportion to the offences committed. Yet, once one realizes that measured over a lifetime repeated occurrences of small inequities in reciprocity may add up to a substantial fitness loss, this becomes more understandable.

In short, contrary to what Gintis et al. (2005) argue, it is not at all obvious that RA is psychologically selfish. We cannot infer from the fact that in RA the lifetime fitness of actors is eventually increased rather than decreased that increasing their own lifetime fitness is what the actors ultimately desire. Assuming otherwise rests on mistaking ultimate for proximate explanation and on confusing instrumentality with functionality.

## 8.4 IS GROUP SELECTION NEEDED FOR SR TO EVOLVE?

Gintis et al. (2005) argue that unlike reciprocal altruists, strong reciprocators engage in costly rewarding and punishing behavior in anonymous one-shot situations when there are no present or future rewards for the reciprocators, and when the recipients are not relatives. Since not just the temporary fitness but also the lifetime fitness of the actor is decreased, SR is evolutionarily altruistic. Standard ultimate explanations such as RA, kin selection (Hamilton 1963) and costly signaling cannot explain this type of altruism. If there had only been selection processes at the level of the individual (e.g. RA and costly signaling), SR could not have evolved. Rather than concluding that SR is maladaptive behavior, as Burnham and Johnson (2005) do, Gintis et al. maintain that it is an adaptation produced by group selection (Fehr and Henrich 2003).

Sometimes proponents of SR seem to argue that group selection was necessary for SR to evolve. But this is contradicted by other statements. Proponents of SR have advanced many different models to explain how it could have evolved. Sethi and Somanathan (2005, p. 243) rightfully speak of a patchwork of models. Not all of these models seem to invoke group selection. In this respect, the very title of Bowles et al. (2003), *Strong reciprocity may evolve with or without group selection*, speaks for itself. Gintis et al. (2001) provide a model of costly signaling, for example, which most commentators would not call an instance of group selection. Fehr and Henrich (2003, pp. 60–61), for example, treat this model as belonging to the camp of standard evolutionary explanations that models of group selection are argued to go beyond. Thus it seems that, *pace* Gintis et al. (2005), models of how SR could have evolved need not invoke selection at a level above that of the individual.

What is the relation between group selection models and standard ultimate explanations like RA and kin selection anyway? Sometimes it seems proponents of SR argue that group selection models identify a way in which, for example, SR could have evolved that is altogether different from the ways that standard ultimate explanations identify. West et al. (2007) distinguish between direct and indirect fitness benefits. Direct fitness benefits obtain if fitness benefits accrue to the actor later in the actor's life. There are indirect fitness benefits if the benefits do not accrue to the actor himself but to genetic relatives of the actor, or, more generally, to other individuals with the same relevant trait. The paradigmatic example of an explanation of evolutionary altruism in terms of indirect fitness benefits is kin selection.

West et al. argue that direct and indirect fitness benefits exhaust the set of all possible explanations of how behavior that momentarily decreases the actor's fitness could have evolved. One way to understand "group selection" is that it presents a third way in which evolutionary altruism could have evolved: evolutionary altruism evolved because groups with evolutionary altruists overrepresented had greater reproductive success than groups with evolutionary altruists underrepresented. On this understanding, "group selection," is strictly analogous to individual selection. In particular, it is assumed that there is group fitness just as much as there is individual fitness. When Gintis et al. (2005) argue that selection on a level above that of the individual must be assumed for SR to have evolved, it seems they have this understanding of "group selection" in mind. West et al.'s argument that direct and indirect fitness benefits provide the only two ways in which evolutionary altruism could have evolved would presuppose that there is only individual selection (or individual selection plus genic selection). This presupposition would have to be discarded to explain how SR could have evolved. SR could only have evolved if there had been group selection and if its force would have been strong enough to counteract the altruism-undermining force of individual selection.[4]

Yet there is overwhelming evidence that this kind of "old group selection" is not what proponents of SR have in mind. The models of group selection put forward by SR theorists do not seem to deny that there must be direct or indirect fitness benefits for cooperation and evolutionary altruism to have evolved. West et al. (2007) correctly observe that at least in some of the evolutionary models advanced, SR could have evolved precisely and only for the reason that in the end there were compensating future fitness benefits for the reciprocator. In Gintis (2000), for example, it is assumed that all members of the group benefit from the behavior of the reciprocator (because it leads to increased productivity, for example, or to a lower rate of group extinction),[5] including the reciprocator himself. In this model, the behavior of the reciprocator can only evolve because the net fitness effect for the reciprocator is positive (and results in a lifetime fitness of the reciprocator that is superior to that of other individuals in the population). Thus, in this model of group selection, SR could have evolved because of direct fitness benefits for reciprocal altruists. Furthermore, if RA is deemed evolutionarily selfish on the ground that it ultimately leads to superior lifetime fitness of reciprocal altruists, then at least in models such as the one presented in Gintis (2000), SR must be regarded as evolutionarily selfish too.

In their response to evolutionary psychologist Price (2008), Gintis et al. (2008) write

We agree with Price that strong reciprocity must have promoted individual fitness, or it could not have evolved. Our contention is that strong reciprocity enhanced relative fitness because groups with a high frequency of altruism survived and prospered at a higher rate than groups with a low frequency of altruism.

Gintis et al. clearly accept here that it is ultimately individual fitness, not group fitness, that must have been promoted for SR to have been able to evolve. The relatively high rate of survival and prospering of groups with

---

[4] Even if there had been only group selection in this "old" sense, individual fitness would be completely derivative from group fitness and it could still be maintained that the individual fitness of reciprocal altruists would exceed that of other types of individuals in the population.

[5] But see Egas and Riedl (2008) for evidence that while altruistic punishment might enhance the level of cooperation, it might lower group performance.

a high frequency of altruism is presented here as one way in which superior individual fitness could be realized. Framed in the terminology introduced earlier, how groups are formed and how they fare compared with others should be functional in reaching that "ultimate end."

This suggests that SR theorists assume "new" rather than "old" group selection. As Okasha (2006) points out, "new group selection" does not call into question the claim that it is ultimately individual fitness, or inclusive fitness, that must have increased to make the evolution of evolutionary altruism possible. Models of new group selection show that the way interactions in populations are clustered in groups might affect individual fitness or inclusive fitness. It has been shown that new group selection is mathematically equivalent to kin selection (see Hamilton 1975; Grafen 1984; Frank 1986). Thus, unlike old group selection, new group selection cannot be presented as if it were totally different from kin selection. If we deem the momentary fitness-reducing behaviors explained in kin selection models evolutionarily altruistic, on the ground that not just the actor's momentary individual fitness is decreased but also the actor's lifetime individual fitness, SR for the very same reason comes out as evolutionarily altruistic in models of new group selection. In both kin selection models and new group selection models it is indirect fitness benefits that make the evolution of evolutionary altruism possible. Once the mathematical equivalence of group selection and kin selection models is appreciated, the debate shifts from the issue of whether some evolutionary process displays either kin selection or group selection to the issue of what, in a particular case, is the more fruitful interpretation of what is, at an abstract level at least, the very same evolutionary process (cf. the exchange between Wilson 2007 and West et al. 2007).

Should we agree with West et al. (2007), then, that there is nothing new under the sun? I disagree. Like standard individualistic ultimate explanations, the group selection explanations of SR also crucially refer either to direct or to indirect fitness benefits for individuals. But these explanations are similar only at this very abstract level of description. At a more concrete level they might differ considerably. The ways in which direct fitness benefits can be recouped can vary greatly. Direct fitness benefits might be recouped by recipients returning the favor, as RA points out. But direct fitness benefits might be recouped also by increased performance of the group that all group members benefit from, as some models of group selection show. Ditto for the pathways via which indirect fitness benefits can be gained. Indirect fitness benefits might be gained by helping relatives, as models of kin selection demonstrate. But indirect fitness benefits might be gained also by interacting only with kindred individuals. In short, it is true that unlike models of old group selection, models of new group selection do not identify, at an abstract level, a third way (next to direct and indirect fitness benefits) in which momentarily fitness-decreasing behavior could have evolved. But at a more concrete level the models of new group selection advanced by proponents of SR point to other pathways than those identified in RA and kin selection (as they are traditionally understood)[6] via which direct and indirect fitness benefits can be obtained.

Once we distinguish between abstract and more concrete levels at which ultimate explanations can be given, it transpires that, contrary to what West et al. (2007) suggest, there is nothing wrong with referring to proximate factors in ultimate explanations of cooperation at a more concrete level. West et al. (2007) criticize SR's proponents for mixing up ultimate and proximate factors, arguing that they confuse things by suggesting that imitation and learning should play an important role in the ultimate explanation of human cooperation. West et al. write: "Learning and imitation are a solution to the proximate not the ultimate problem. … The ultimate question is why is cooperation and punishment favoured in the first place" (ibid., p. 426). West et al. are right, of course, that ultimate and proximate explanations should not be mixed up. But I don't think proponents of SR are mixing them up here. West et al. seem to assume that in answering the ultimate question of why some behavior could have evolved in the first place only reference to ultimate causes (notably natural selection favoring behavior with superior fitness) is admissible. Reference to any proximate cause or mechanism should be shunned. This assumption seems unwarranted. If imitation and learning actually did play an important role in the evolution of human cooperation then it is hard to see why reference to them should be inadmissible in the ultimate explanation of why human cooperation could have evolved in the first place (see Marchionni and Vromen 2009).

The situation here is similar to that of group selection. West et al. are right to argue, and proponents of SR seem to agree, that models of group selection cannot identify a third way in which cooperation could have evolved, next to the occurrence of direct or indirect fitness benefits. Models stressing the role of social transmission should similarly be

---

[6] "Kin selection" and "reciprocal altruism" are sometimes interpreted more broadly, as encompassing helping nonrelatives and nondyadic interactions. If they are understood that broadly, new group selection models identify pathways other than helping relatives and dyadic interactions within the rubrics of kin selection and reciprocal altruism.

rejected unless they can show that eventually in one way or another direct and/or indirect fitness benefits are realized. But this is a condition that all acceptable models have to meet at the abstract level. At a more concrete level, the ways in which direct and/or indirect fitness benefits are realized might crucially involve "group selection" (in the sense of group competition) and social transmission. If so, there is no good reason to shun reference to them in ultimate explanation. Indeed, if group competition and social transmission actually have been important factors in the evolution of human cooperation, then there is a good reason to demand that they should be referred to in ultimate explanation. It is interesting to see that West et al. do not deny that group competition and social transmission might have been potent forces in the evolution of human cooperation (see also West et al. 2009). West et al. (2007, p. 427) observe that what might be special about the evolution of cooperation in humans is the interaction of ultimate and proximate mechanisms that might have led to a specific evolutionary dynamic.[7] Thus, although the models of the evolution of SR at an abstract level of description must refer to offsetting direct and/or indirect fitness benefits for strong reciprocators, at a more concrete level they might be informative about the sorts of processes that actually went into producing these direct and/or indirect fitness benefits. As Boyd (2006) argues, proponents of SR believe it is at this more concrete level that the really interesting questions should be posed and answered.

I conclude that SR could have evolved without group selection. Furthermore, the models of group selection that SR theorists have put forward do not identify a third way, next to direct and indirect fitness benefits, in which cooperation and evolutionary altruism could have evolved. In some of these models direct fitness benefits are crucially referred to. If RA is deemed evolutionarily selfish on the ground that lifetime individual fitness is increased, then SR also comes out as evolutionarily selfish in these models. In other models, indirect fitness benefits play a crucial role. If kin selection is deemed evolutionarily altruistic on the ground that lifetime individual fitness is decreased, then in these models SR also appears as evolutionarily altruistic. In short, the contribution of these models of "new" group selection does not lie in identifying a new way in which cooperation could have evolved at an abstract level of description. But the models point at pathways, other than RA and kin selection (interpreted

---

[7] In Vromen (2009) I call the proximate mechanisms working way back in time *then-working* proximate causes, to distinguish them from now-working proximate causes that produce the behavior to be explained. I also argue there that then-working proximate causes also determined the then-available strategy set.

narrowly), via which cooperation could have evolved at a more concrete level of description.

### 8.5 IS SR PSYCHOLOGICALLY ALTRUISTIC?

Proponents of SR argue that unlike reciprocal altruists, strong reciprocators are led by non-self-regarding motives. If strong reciprocators had self-regarding motives, they argue, it would be inexplicable that they reward norm followers and punish norm violators even at considerable personal cost and even in anonymous one-shot games when they cannot plausibly expect that these costs are recouped later. Proponents of SR also argue that there is a need to incorporate social preferences, such as a taste for fairness or the desire that norms of cooperation be complied with, as separate arguments in utility functions. All this seems to suggest that proponents of SR believe that SR is not psychologically selfish.

But drawing this conclusion might be overhasty. Behavior that is led by the desire to restore justice rather than by a desire to maximize one's own material rewards, for example, might be psychologically selfish. If the desire to restore justice is not an ultimate desire, but an instrumental one produced in a process of practical reasoning that starts with the ultimate desire to maximize one's own psychic satisfaction, the behavior is psychologically selfish. What determines whether or not some behavior is psychologically selfish is whether or not the ultimate desire is self-regarding. Whether or not instrumental desires, which are formed on the basis of ultimate desires and beliefs, are self-regarding is irrelevant. Maximizing one's own psychic satisfaction is clearly a self-regarding desire, whereas maximizing the well-being of someone else clearly is an other-regarding desire. But if one wants to maximize someone else's well-being because one wants to feel good and because one believes that maximizing someone else's well-being is the best way to feel good, then the ultimate desire is self-regarding and the behavior resulting from it is psychologically selfish.

On the basis of the results of PET scans, de Quervain et al. (2004) argue that altruistic punishment is psychologically selfish. De Quervain et al. found that the caudate nucleus (as part of the dorsal striatum) is especially active in altruistic punishers. Since earlier studies done by Wolfram Schultz and others (see Schultz and Romo 1988; Schultz 2000) had shown that the caudate nucleus is implicated in goal-directed behavior, de Quervain et al. conclude from this that in their punishing behavior strong reciprocators are led by an anticipation of rewards. Strong

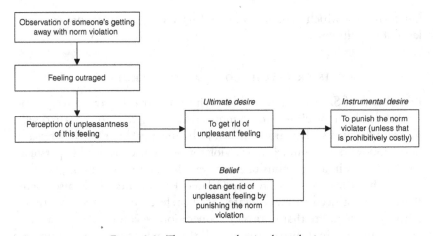

Figure 8.3. The outrage-reduction hypothesis.
*Note:* This depiction of the outrage-reduction hypothesis is based partly on Clavien and Klein's (2010) discussion of the de Quervain et al. findings. Unlike Clavien and Klein (2010) and Kitcher (2010), I do not see a need to distinguish evolutionary altruism and psychological altruism from behavioral altruism.

reciprocators punish norm violators because they expect to reap psychic satisfaction from doing so. They "feel bad if they observe that norm violations are not punished, and they seem to feel relief and satisfaction if justice is established. Many languages even have proverbs indicating such feelings, for example, 'Revenge is sweet'" (de Quervain et al. 2004, p. 1254). As it is the actor's own hedonic rewards that are ultimately driving altruistic punishment, de Quervain et al. argue, altruistic punishment is psychologically selfish (ibid., p. 1257).

Schematically, it seems that the psychological process altruistic punishers go through according to de Quervain et al. (2004) can be depicted as in Figure 8.3.

Altruistic punishers feel outraged if they observe that norm violators get away with their behavior. Their ultimate desire is to get rid of this negative feeling. They believe that they will experience a compensating positive feeling by taking revenge on the norm violator. This is how they develop, via a process of practical reasoning, the instrumental desire to take revenge. Because the ultimate desire is a self-regarding one, altruistic punishment is psychologically selfish.

Contrary to what de Quervain et al. (2004) argue, however, it is far from clear that this outrage-reduction hypothesis is suggested (or even supported) by the results of their PET scans. As Poldrack (2006) argues, the reverse inference that de Quervain et al. make from earlier brain scans (and neurobiological studies) is fallacious. The fact that particular brain areas (such as the caudate nucleus) have been observed to be activated in particular mental tasks and processes (such as goal-directed behavior) does not imply that the same mental tasks and processes occur whenever these brain areas are active. The simple reason is that the brain areas might be activated also if other mental tasks and processes occur. Our current understanding of specialization in the brain suggests that the sort of strong selectivity simply does not exist that would warrant making reverse inferences. More importantly, de Quervain et al. seemed to be mistaken in taking earlier studies to show that it is via a process of practical reasoning that behavior guided by expectations of rewards is connected with the activation of the caudate nucleus. One could say that these earlier studies by Schultz and others do show that the expectations of future rewards play a role in processes of valuation, but not that these expectations play the role of beliefs in processes of practical reasoning.

Based on the pioneering work of Schultz and others, neurobiologists now believe that valuation in the brain is a kind of reinforcement learning. At the level of dopaminergic neurons, actors learn the subjective values things have for them by coding how strongly actual rewards differ from expectations. As long as actual rewards do not deviate from their present subjective valuations, the latter are reinforced. Actors only change their subjective values if actual rewards differ from their expectation. These studies do not posit that actors act upon expectations about future rewards in processes of practical reasoning. The mode of behavior, the way in which actors make their decisions, might rather be instinctive, habitual or directly enticed by emotions. Thus, when a monkey instinctively reaches out to grab the squirt of orange juice that is offered to him, the monkey is implicitly and unconsciously acting on the subjective value the monkey has learned to attach to the squirt. The latter can be called an expected value. But the behavior is not consciously guided by expectations of future rewards.

It seems that de Quervain et al.'s PET-scan findings are compatible with a host of other hypotheses. One such hypothesis, I submit, is reminiscent of Trivers' notion of moralistic aggression and can be called the moral indignation–spite hypothesis (see Figure 8.4).

In a sense, the moral indignation–spite hypothesis is a simpler shortcut version of the more elaborate outrage-reduction hypothesis. The moral indignation–spite hypothesis asserts that the causal link between feeling of outrage (or moral indignation) and the desire to harm (or punish) the

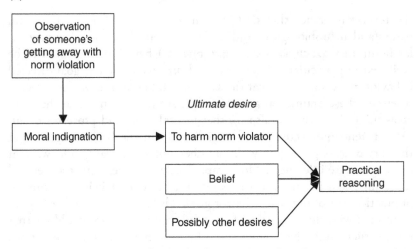

Figure 8.4. The moral indignation–spite hypothesis.

norm violator is not mediated by the desire to get rid of the unpleasant feeling. Moral indignation or moral outrage directly evokes the desire in altruistic punishers to harm norm violators by taking revenge, according to the moral indignation–spite hypothesis. Altruistic punishers implicitly expect others also to comply with the prevailing social norms. This expectation is violated by norm violators (and this element of "surprise" might explain the increased activation of the caudate nucleus) and that makes them angry.[8] This negative punitive emotion has a direct action tendency (Dubreuil 2010): it prompts altruistic punishers to take revenge.

As practical reasoning only starts here after the desire to inflict harm on the norm violator is formed, the desire to inflict harm is an ultimate one. Is this ultimate desire self-regarding? Clavien and Klein (2010), who offer an interpretation similar to the moral indignation–spite hypothesis, argue that it is. Clavien and Klein argue that the desire for revenge is self-regarding. Thus, since the ultimate desire is deemed self-regarding, altruistic punishment once again comes out as psychologically selfish. I am not convinced that the desire for revenge is self-regarding, however. No doubt the desire for revenge can sometimes be called self-regarding,

---

[8] In the following discussion, I simply assume that feeling outrage, feeling anger and feeling moral indignation amount to much the same thing. See Dubreuil (2010) for a finer-grained discussion of differences between these feelings. I also assume that the desire to take revenge and the desire to harm or punish norm violators amount to the same thing. For other purposes, finer-grained distinctions might be needed here as well.

for example when the target for revenge deliberately inflicted personal harm on the actor. But in the case of altruistic punishment it is the feeling that the norm violator did something *morally* reprehensible that produces the desire to punish the norm violator (see Sanfey et al. 2003). As de Quervain et al. suggest, altruistic punishers feel that justice is being violated by the norm violators. The expectations of the actor here are normative rather than empirical. They pertain to what others ought to do rather than to what they will do. In the case of norm violation, the actor does not blame himself for having had incorrect expectations, but the perpetrator for not living up to the actor's expectations.

Furthermore, the desire to punish norm violators is directed at doing something to others rather than at accomplishing something for oneself. In this sense, the moral indignation–spite hypothesis is similar to Batson et al.'s (1991) empathy–altruism hypothesis. This hypothesis is perhaps the best-known contemporary version of the view that psychological altruism does exist. It asserts that the distress observed in others might engender high empathy in an actor, which in turn can evoke a desire in the actor to contribute to the well-being of the others. The desire to contribute to the well-being of the others is an other-regarding one, and according to Batson et al. it is not produced in a process of practical reasoning by a self-regarding ultimate desire plus a suitable belief. Instead it is directly caused by the actor's high empathy. That is why the other-regarding desire is ultimate. Hence the behavior described in the empathy–altruism hypothesis qualifies as psychological altruism. On the moral indignation–spite hypothesis, the desire to punish norm violators is likewise directly produced by the moral indignation of the actor, which in turn is engendered by observing a norm violation. Unlike the desire to contribute to the well-being of others, the desire to harm others is clearly not an altruistic desire. It rather is a spiteful desire. But like the altruistic desire to contribute to the well-being of others, the spiteful desire to harm others is other-regarding, not self-regarding. Hence what we should conclude, I think, is that on the moral indignation–spite hypothesis altruistic punishment is neither psychologically selfish nor altruistic.

## 8.6 CAN BRINGING IN PSYCHOLOGICAL AND EVOLUTIONARY EVIDENCE SETTLE THE ISSUE?

The available neurobiological evidence (as reported in de Quervain et al. 2004) falls short of conclusively favoring the more elaborate psychologically selfish rendering of altruistic punishment over the simpler psychologically nonselfish moral indignation–spite hypothesis. Both theses, I have argued, are compatible with the evidence provided by de Quervain et al. (2004). But perhaps there are other sources of evidence that we can draw on to settle the issue. Here, I now want to argue, we can learn from the debate over psychological altruism between Batson, with his empathy–altruism hypothesis, on the one hand and more skeptical commentators such as Sober and Wilson (1998), Stich (2007) and Stich et al. (2010) on the other. I argue that the message we can take away from this debate, however, is mainly negative: adding psychological and evolutionary evidence does not conclusively settle the issue either.

In several careful and cleverly conducted experiments Batson and his colleagues have attempted to rebut many of the specific explanations of helping behavior in terms of psychological selfishness. One of these is the aversive-arousal reduction hypothesis: helpful actors do not ultimately desire the improved well-being of others, but the elimination of their own unpleasant feelings of distress that observing others in distress (and empathizing with them) might bring along. By giving distressed experimental subjects opportunities to get rid of their unpleasant feelings other than by helping others in distress, Batson and his colleagues tried to establish that it is really the ultimate altruistic desire for the well-being of others and not some ultimate hedonistic desire that is driving the helping behavior. In one experiment, Batson and his associates offered subjects an escape option. Subjects could walk away from the sight of others in distress. Batson et al. reasoned that if the empathy–altruism hypothesis is true, subjects would still choose to help the others. But if the aversive-arousal reduction hypothesis is true, subjects would choose the escape option as a cheaper solution to getting rid of their unpleasant feelings. What they found is that the vast majority of the subjects still chose to help others, which Batson et al. regarded as psychological evidence supporting their empathy–altruism hypothesis.

It is easy to think of a similar sort of experiment to test the moral indignation–spite hypothesis against the rival outrage-reduction hypothesis advanced by de Quervain et al. (2004): altruistic punishers punish norm violators because they thereby expect to reduce (or to altogether get rid of) the negative feelings that the observation of as-yet-unpunished, intentional norm violators induces in them. Give altruistic punishers the easy escape option of walking away and see what they do next. If most of them go for the escape option and if their desire to harm norm violators declines, this provides evidence in favor of the rival hypothesis. If not, evidence is found in favor of the moral indignation–spite hypothesis.

As Stich et al. (2010) convincingly point out, however, Batson et al.'s interpretation rests on a dubious assumption. Batson et al. assume that if subjects are given an easy escape option in which they can forget about the misery of others, their own personal distress will be reduced. As the old adage has it, "out of sight, out of mind." But it is dubious that this is true of high-empathy subjects. For high-empathy subjects taking more physical distance will probably not mean taking more psychological distance (i.e. they will continue to care about the well-being of the people in distress). If the psychological distance does not increase, selfish aversive-arousal reducers will still believe that the only way to reduce their own distress is to reduce that of others.[9] Thus the finding of Batson et al., that the desire to help others does not decline, does not really discriminate between the two hypotheses.[10] It is not just that sometimes experimental findings fail to clearly favor one hypothesis over the other; it seems that, with some ingenuity, more sophisticated new rival hypotheses could always be concocted, if some specific alternative has been disconfirmed (but see Rosas 2002 on the limited persuasiveness of this "open-set problem").

I conclude that psychological evidence is not likely to be able to settle the issue of which interpretation of the psychological process underlying altruistic punishment is the right one. This is also argued by Sober and Wilson (1998) with respect to the more general issue of whether or not psychological altruism exists. Sober and Wilson argue that evolutionary considerations might provide a more promising avenue for settling the issue. Their basic idea is that ultimate causes never produce behavior directly, but always only indirectly, via the production of proximate causes. Sober and Wilson argue that evolutionary considerations can be put to use to find out what in a specific case is the most plausible hypothesis: psychological selfishness, psychological altruism or motivational pluralism. They discuss three such evolutionary considerations: availability (the proximate cause must have been available in ancient populations),

---

[9] See also Cialdini et al.'s (1997) nonaltruistic reinterpretation of the data supporting Batson's empathy–altruism hypothesis. Cialdini et al. argue that benevolence towards others with whom we have developed close attachments is not the result of our feeling more empathic concern for them, but the result of our feeling more at one with them: we then have a sense of "inclusive self" with them (similar to "inclusive fitness").

[10] Stich et al. (2010) argue that the results reported in Stocks et al. (2009) do seem to support Batson's empathy–altruism hypotheses in comparison to the aversive–arousal reduction hypothesis. But other studies, such as Dana et al. (2006), show that subjects choose cheap escape options when they are offered to them, suggesting that they do not really care about fairness or justice.

reliability (the proximate cause must produce the behavior that is most functional in a reliable way) and efficiency (the proximate cause must produce the behavior that is most functional efficiently). In the particular example Sober and Wilson discuss, that of parental care, the consideration of reliability is doing most of the work.

The issue Sober and Wilson address is whether the ultimate desire that gives rise to help offered by parents to their children is other-directed or self-directed (or both, in case of motivational pluralism). If the ultimate desire of parents is solely to improve the welfare of their children, the ultimate desire is other-directed and hence psychological altruism exists. By contrast, if the ultimate desire of parents is solely to maximize their own pleasure and minimize their own pain, it is self-directed and hence the helping behavior is a case of psychological selfishness. The practical reasoning process on the second "selfishness" hypothesis is more complex and elaborate than the process on the first "altruism" hypothesis. In particular, there is one more mediating belief involved in the second than in the first process, namely the belief that the ultimate desire to maximize one's own pleasure and minimize one's own pain is best served by helping one's children when they are in need of help. Sober and Wilson note that people might err in the formation of their beliefs. This makes the provision of parental help less reliable in the second process than it is in the first. If parents come to believe that their hedonic rewards are better served by spending time on leisurely activities rather than by helping their children, for example, parental help will be less forthcoming on the second than on the first hypothesis. On the basis of this evolutionary consideration, Sober and Wilson conclude that psychological selfishness is less likely to have evolved than psychological altruism (and, in particular, than motivational pluralism).

This has implications for the difference between the moral indignation–spite hypothesis and the de Quervain et al. (2004) interpretation of altruistic punishment. Compared with the moral indignation–spite hypothesis, the practical reasoning process in the de Quervain et al. (2004) interpretation is more complex, in that it posits the existence of one more mediating belief. On the de Quervain et al. (2004) interpretation, if actors do not form the mediating belief that they will get rid of their unpleasant feeling by punishing norm violators, the desire to punish norm violators will not arise. By contrast, the desire to do so will arise on the moral indignation–spite hypothesis, no matter what beliefs (if any) actors might have about the relation between the feeling of moral indignation and punishing norm violators. Since altruistic punishment is more reliable on the moral indignation–spite hypothesis, it is more likely that the psychological mechanism specified by the hypothesis has evolved.

Should we conclude that this provides conclusive evolutionary evidence in favor of the moral indignation–spite hypothesis? Stich (2007) is not persuaded by Sober and Wilson's evolutionary argument. Stich argues that Sober and Wilson assume that the relevant sorts of beliefs connecting ultimate and instrumental desires are empirical, inferentially integrated beliefs that can be generated and removed by inferential processes in which other beliefs and (possibly misleading) evidence play a key role. Stich goes on to argue that not all belief-like states are like that. Some such states might be "subdoxastic states" that are "stickier" than inferentially integrated beliefs. Given the obvious evolutionary importance of parental care, it might well be a subdoxastic state that accounts for the crucial mediating belief in the "selfishness" hypothesis, that the ultimate desire to maximize one's own pleasure and minimize one's own pain is best served by helping one's children when they are in need of help. If it is, then, especially if one believes that ultimate desires tend to be more stable than instrumental ones,[11] the psychological mechanism involved in the more elaborate psychological selfishness interpretation might be just as reliable as the simpler psychological altruism interpretation. The ultimate desire and the relevant belief then would come in, as it were, a fixed, stable package, more or less guaranteeing that the (supposedly) relevant instrumental desire is produced. Similarly, if Trivers is right that upholding schemes of cooperation is of vital importance, the mediating belief that the unpleasant feelings evoked by observing norm violations will be relieved by punishing the norm violators might also be a subdoxastic state. The psychological mechanism described by the outrage-reduction hypothesis would be as reliable at producing punishing behavior as that of the moral indignation–spite hypothesis.

Once again the conclusion seems to be that bringing in extra sorts of evidence (this time from evolutionary considerations) does not settle the issue of whether altruistic punishment is psychologically altruistic. It seems that the inconclusiveness cannot be eliminated completely. It is not so clear, however, how troublesome this is. No matter whether the ultimate desire driving altruistic punishment is an other-directed or a stable self-directed one (in combination with "sticky" subdoxastic states), the

---

[11] Batson et al. (1991) make this assumption implicitly when they argue that if people find out that some desire is not instrumental to satisfying an ultimate desire, they are likely to continue pursuing the ultimate desire while ceasing to pursue the instrumental one.

same patterns of behavior are reliably produced. Whether or not "altruistic punishment" denotes a reliable and robust pattern of behavior is more important and interesting than whether or not the practical reasoning processes altruistic punishers go through involve one more error-free step.

Rather than contrasting SR with RA with respect to both evolutionary altruism and psychological altruism, I think proponents of SR can serve their cause better by further explorations of the conditions under which people tend to punish norm violators at a considerable personal cost. Obvious variables are the magnitude of the personal costs, the perceived seriousness and wrongness of the norm violations, whether or not there are others around who could do the punishing, and the chances that the costs will be paid back later. Some of these variables may play an important role even if they are not consciously registered by the actors. As Haley and Fessler (2005) and Hagen and Hammerstein (2006) point out, for example, having the feeling that one is being watched can be triggered by subtle environmental cues even in anonymous one-shot games. Doing more work on these issues seems to be more productive than highlighting allegedly unique features of SR.

## 8.7 CONCLUSION

SR and RA might be much more alike than proponents and opponents of SR suggest. If RA is regarded as evolutionarily selfish on the ground that RA leads to superior lifetime fitness, then on the very same ground SR comes out as evolutionarily selfish too in at least some of the ultimate explanations given of SR. Proponents of SR have advanced many models of how it could have evolved. Only some of them invoke group selection. What is more, the models of group selection advanced do not challenge the standard "individualist" dictum that momentarily fitness-decreasing behavior could not have evolved without compensating direct or indirect fitness benefits. Thus with respect to the issue of whether evolutionary altruism or selfishness is at stake, it seems SR and RA are in the same position. The situation with respect to the debate over whether SR and RA are psychologically altruistic is much less clear. If we assume, as is commonly done in debates over psychological altruism, that the behaviors associated with SR and RA are derived by actors in processes of practical reasoning from some particular ultimate desire, it seems that inconclusiveness reigns. It seems that neither neuroimaging, nor psychological, nor evolutionary evidence (nor all three in combination) can settle the issue. What is clear, though, is that, contrary to what proponents of SR sometimes argue, RA need not be psychologically selfish. As Trivers (1971) already discussed under the rubric of moralistic aggression, prosocial emotions (rather than self-serving calculation) might induce other-directed desires. Likewise, contrary to what de Quervain et al. argue, altruistic punishment need not be psychologically selfish either. A similar proximate psychological mechanism as Trivers suggested for RA, which I dubbed the moral indignation–spite hypothesis, might also underlie altruistic punishment.

The debate over whether SR and RA are ultimately selfish seems to divert attention from more important issues. One such issue is how reliable and robust the behavior patterns posited by SR actually are. How sensitive is the conditional rewarding and punishing behavior with respect to changes in the magnitude of personal costs and in the perceived effectiveness of the rewarding and punishing behavior, for example? Another important issue pertains to the pathways via which SR could have actually evolved. Contrary to what some proponents of SR seem to suggest, kin selection and group selection are mathematically equivalent. But beneath this highly abstract level of equivalence, at a more concrete level of analysis, several sorts of evolutionary processes might have been going on that are profoundly different and that might not all be equally plausible. At a more concrete level, preferentially favoring close relatives, which kin selection on a "literal" translation amounts to, is quite unlike cultural group competition (with its emphasis on conformist social transmission), for example. If the aim of ultimate explanations of the evolution of SR is not to demonstrate SR's survival value, but to give a plausible account of how SR actually evolved, then such differences really matter. Thus, although from a highly abstract point of view the models provided by proponents of SR are old wine in new bottles, as attempts to sketch plausible evolutionary scenarios of how SR could actually have evolved their efforts should be welcomed.

## REFERENCES

Batson, C. D., Batson, J. G., Slingsby, J. K., Harrell, K. L., Peekna, H. M. and Todd, R. M. 1991. "Empathic joy and the empathy-altruism hypothesis," *Journal of Personality and Social Psychology* 61(3): 413–26.

Binmore, K. 2004. "Reciprocity and the social contract," *Politics, Philosophy & Economics* 3(1): 5–35.

Bowles, S., Fehr, E. and Gintis, H. 2003. "Strong reciprocity may evolve with or without group selection," *Theoretical Primatology Project Newsletter* 1(12) (supplement).

Boyd, R. 2006. "The puzzle of human sociality," *Science* 314(5805): 1555–6.

Burnham, T. C. and Johnson, D. D. P. 2005. "The biological and evolutionary logic of human cooperation," *Analyse & Kritik* 27(2): 113–35.

Cialdini, R. B., Brown, S. L., Lewis, B. P., Luce, C. and Neuberg, S. L. 1997. "Reinterpreting the empathy–altruism relationship: When one into one equals oneness," *Journal of Personality and Social Psychology* 73(3): 481–94.

Clavien, C. and Klein, R. A. 2010. "Eager for fairness or for revenge? Psychological altruism in economics," *Economics and Philosophy* 26(3): 267–90.

Dana, J., Cain, D. M. and Dawes, R. M. 2006. "What you don't know won't hurt me: Costly but quiet exit in dictator games," *Organizational Behavior and Human Decision Processes* 100(2): 193–201.

de Quervain, D. J.-F., Fischbacher, U., Treyer, V., Schellhammer, M., Schnyder, U., Buck, A. and Fehr, E. 2004. "The neural basis of altruistic punishment," *Science* 305: 1254–58.

Dubreuil, B. 2010. "Punitive emotions and norm violations," *Philosophical Explorations* 13(1): 35–50.

Egas, M. and Riedl, A. 2008. "The economics of altruistic punishment and the maintenance of cooperation," *Proceedings of the Royal Society B – Biological Sciences* 275(1637): 871–8.

Fehr, E. and Henrich, J. 2003. "Is strong reciprocity a maladaption?" in Hammerstein (ed.), 55–82.

Frank, S. A. 1986. "Hierarchical selection theory and sex ratios. I. General solutions for structured populations," *Theoretical Population Biology* 29: 312–42.

Gigerenzer, G. 2007. *Gut Feelings: The Intelligence of the Unconscious*, New York: Viking Press.

Gintis, H. 2000. *Game Theory Evolving*, Princeton University Press.

Gintis, H., Bowles, S., Boyd, R. T. and Fehr, E. 2005. "Moral sentiments and material interests: Origins, evidence, and consequences," in Gintis, H., Bowles, S., Boyd, R. T. and Fehr, E. (eds.), *Moral Sentiments and Material Interests: The Foundations of Cooperation in Economic Life*, Cambridge, MA: MIT Press, 3–39.

Gintis, H., Henrich, J., Bowles, S., Boyd, R. and Fehr, E. 2008. "Strong reciprocity and the roots of human morality," *Social Justice Research* 21(2): 241–53.

Gintis, H., Smith, E. A. and Bowles, S. 2001. "Costly signaling and cooperation," *Journal of Theoretical Biology* 213: 103–19.

Grafen, A. 1984. "Natural selection, kin selection and group selection," in Krebs, J. R. and Davies, N. B. (eds.), *Behavioural Ecology. An Evolutionary Approach*, Oxford: Blackwell Scientific, 62–84.

Hagen, E. H. and Hammerstein, P. 2006. "Game theory and human evolution: A critique of some recent interpretations of experimental games," *Theoretical Population Biology* 69(3): 339–48.

Haley, K. J. and Fessler, D. M. T. 2005. "Nobody's watching? Subtle cues affect generosity in an anonymous economic game," *Evolution and Human Behavior* 26: 245–56.

Hamilton, W. D. 1963. "The evolution of altruistic behavior," *American Naturalist* 97: 354–6.

1975. "Innate social aptitudes in man: An approach from evolutionary genetics," in Fox, R. (ed.), *Biosocial Anthropology*, New York: Wiley, 133–55.

Hammerstein, P. (ed.) 2003. *Genetic and Cultural Evolution of Cooperation*, Cambridge, MA: MIT Press.

Kitcher, P. 2010. "Varieties of altruism," *Economics and Philosophy* 26(2): 121–48.

Marchionni, C. and Vromen, J. 2009. "The ultimate/proximate distinction in recent accounts of human cooperation," *Tijdschrift voor Filosofie* 71(1): 87–117.

Mayr, E. 1961. "Cause and effect in biology," *Science* 134(3489): 1501–6.

Okasha, S. 2006. *Evolution and the Levels of Selection*, Oxford University Press.

Poldrack, R. A. 2006. "Can cognitive processes be inferred from neuroimaging data?" *Trends in Cognitive Sciences* 20(20): 59–63.

Price, M. E. 2008. "The resurrection of group selection as a theory of human cooperation," *Social Justice Research* 21(2): 228–40.

Rosas, A. 2002. "Psychological and evolutionary evidence for altruism," *Biology and Philosophy* 17(1): 93–107.

Sanfey, A. G., Rilling, J. K., Aronson, J. A., Nystrom, L. E. and Cohen, J. D. 2003. "The neural basis of economic decision-making in the ultimatum game," *Science* 300(5626): 1755–8.

Schultz, W. 2000. "Multiple reward signals in the brain," *Nature Reviews Neuroscience* 1: 199–207.

Schultz, W. and Romo, R. 1988. "Neuronal activity in the monkey striatum during the initiation of movements," *Experimental Brain Research* 71: 431–6.

Sethi, R. and Somanathan, E. 2005. "Norm compliance and strong reciprocity," in Gintis, H., Bowles, S., Boyd, R. T. and Fehr, E. (eds.) *Moral Sentiments and Material Interests: The Foundations of Cooperation in Economic Life*, Cambridge, MA: MIT Press, 229–251.

Sober, E. and Wilson, D. S. 1998. *Unto Others: The Evolution and Psychology of Unselfish Behavior*, Cambridge, MA: Harvard University Press.

Stevens, J. R. and Hauser, M. D. 2004. "Why be nice? Psychological constraints on the evolution of cooperation," *Trends in Cognitive Sciences* 8(2): 60–5.

Stich, S. 2007. "Evolution, altruism and cognitive architecture: A critique of Sober and Wilson's argument for psychological altruism," *Biology and Philosophy* 22: 267–81.

Stich, S., Doris, J. M. and Roedder, E. 2010. "Altruism," in Moral Psychology Research Group (ed.), *The Oxford Handbook of Moral Psychology*, Oxford University Press, 147–205.

Stocks, E., Lishner, D. and Decker, S. 2009. "Altruism or psychological escape: Why does empathy promote prosocial behavior?" *European Journal of Social Psychology* 39: 649–65.

Tinbergen, N. 1963. "On aims and methods of ethology," *Zeitschrift für Tierpsychologie* 20: 410–33.

Trivers, R. 1971. "The evolution of reciprocal altruism," *Quarterly Review of Biology* 46(1): 35–57.

Vromen, J. J. 2009. "Advancing evolutionary explanations in economics: The limited usefulness of Tinbergen's four-question classification," in Ross, D. and Kincaid, H. (eds.), *The Oxford Handbook of Philosophy of Economics*, Oxford University Press, 337–67.

West, S. A., El Mouden, C. and Gardner, A. 2009. "Social evolution theory and its application to the evolution of cooperation in humans," working paper.

West, S. A., Griffin, A. S. and Gardner, A. 2007. "Social semantics: Altruism, cooperation, mutualism, strong reciprocity and group selection," *Journal of Evolutionary Biology* 20(2): 415–32.

Wilson, D. S. 2007. "Social semantics: Toward a genuine pluralism in the study of social behaviour," *Journal of Evolutionary Biology* 21(1): 368–73.

# CHAPTER 9

# Team reasoning, framing and cooperation

## Natalie Gold

Cooperation is puzzling for orthodox game theory because, although cooperation is arguably rational and a substantial number of people cooperate in real life, game theory cannot explain why cooperation is rational nor predict that rational players will cooperate in one-shot games. Solving these puzzles of cooperation is one of the motivations behind the development of a body of decision-theoretic literature on *team agency*.[1] This seeks to extend standard game theory, where each individual asks separately 'What should I do?', to allow teams of individuals to count as agents and for players to ask the question 'What should we do?' This leads to *team reasoning*, a distinctive mode of reasoning that is used by members of teams, and which may result in cooperative actions. The basic idea is that, when an individual reasons as a member of a team, she considers which *combination* of actions by members of the team would best promote the team's objective, and then performs her part of that combination. The rationality of each individual's action derives from the rationality of the joint action of the team.

One can distinguish between the generic idea of team reasoning and specific versions proposed by different people. The most fully developed theories of team reasoning are those of Michael Bacharach (1999, 2006) and Robert Sugden (1993, 2003). They differ in important ways regarding: what happens when there is not common knowledge of group membership, what the group agent should take as its goals, and how group agency comes about. In this chapter, I will explore some of the ramifications of each of these differences.

Thanks to Samir Okasha for comments and discussion, and to audiences at the University of British Columbia and the CUNY Seminar in Logic and Games, where preliminary versions of some of this material were presented. This work was supported by a British Academy Small Research Grant, which I gratefully acknowledge.

[1] See Hodgson 1967; Regan 1980; Gilbert 1987; Hurley 1989; Sugden 1993, 2003; Hollis 1998; Bacharach 1999, 2006; and Anderson 2001.

After showing how a basic version of team reasoning with common knowledge solves problems of cooperation (Sections 9.1 and 9.2), by comparing what Bacharach and Sugden advocate that players should do in the prisoner's dilemma when there is not common knowledge of group identification, I will clarify the concept of a group goal in each of their theories (Section 9.3). By comparing how group agency comes about, and the role of framing in each of Bacharach's and Sugden's theories, I will offer some insights into framing in decision making in general (Section 9.4). Finally, I compare team reasoning with payoff-transformation theories of cooperation (Section 9.5).

One can draw an analogy between rational choice theory and natural selection. While the rational agent of decision theory maximizes utility, evolutionary processes maximize fitness. There is also an analogy between team reasoning and group selection: in team reasoning, the team (or the team utility function) is an additional primitive to those found in standard decision theory, while theories of group selection add groups to the ontology of evolutionary biology. At points throughout the chapter I will draw the analogy. However, note that it is only a structural analogy, suggesting a common mathematical framework, and, as I show in Section 9.5, the truth or falsity of group selection has no bearing on the truth or falsity of team reasoning and vice versa.

## 9.1 THE PUZZLE OF COOPERATION: STAG HUNTS AND PRISONER'S DILEMMAS

Common models of cooperation are examples of what Michael Bacharach (2006) called *games with scope for common gain*, or *games with scope* for short. In games with scope, there exists a Nash equilibrium whose outcome could be improved upon for both players if at least one of them plays a different strategy. (Kollock (1998a) gives this as the definition of a *social dilemma*.) Games with scope offer the possibility of mutual benefit and, hence, we may think of them as offering the possibility of cooperation. Two games that are closely identified with the problem of cooperation are the prisoner's dilemma and the stag hunt.[2]

The prisoner's dilemma, whose generic form is illustrated in Figure 9.1, has been the *locus classicus* in the study of cooperation (e.g. Taylor 1987).

---

[2] Games with scope also include co-ordination games where the equilibria have different payoffs. One example is the game of Hi-Lo, which has also attracted attention in the team-reasoning literature, e.g. Sugden 1993; Bacharach 2006; Gold and Sugden 2007a; Bardsley 2007; Gold and Sugden 2007b.

|  |  | Player 2 | |
|---|---|---|---|
|  |  | *Cooperate* | *Defect* |
| Player 1 | *Cooperate* | b, b | d, a |
|  | *Defect* | a, d | c, c |

$a > b > c > d$

$b > (a + d)/2$

Figure 9.1. The prisoner's dilemma.

In specifying the payoffs of this game, we require only that they are symmetrical between the players and that they satisfy two inequalities. The inequality $a > b > c > d$ encapsulates the central features of the prisoner's dilemma: for each player, the best outcome is that in which he chooses *defect* and his opponent chooses *cooperate*; the outcome in which both choose *cooperate* is ranked second; the outcome in which both choose *defect* is ranked third; and the outcome in which he chooses *cooperate* and his opponent chooses *defect* is the worst of all. The inequality $b > (a + d)/2$ stipulates that each player prefers a situation in which both players choose *cooperate* to one in which one player chooses *cooperate* and the other chooses *defect*, each player being equally likely to be the free rider. This condition is usually treated as a defining feature of the prisoner's dilemma.

For each player, *defect* strictly dominates *cooperate*. Each individual player can reason to the conclusion, 'The action that gives the best result *for me* is *defect*'. Thus, in its explanatory form, conventional game theory predicts that both players will choose *defect*. In its normative form, it recommends *defect* to both players. Yet both would be better off if each chose *cooperate* instead of *defect*. Thus, each player can also reason to the conclusion: 'The pair of actions that gives the best result *for us* is not (*defect*, *defect*)'.[3]

The game-theoretic puzzle of cooperation is isomorphic to the biological problem of its evolution, often referred to by biologists as the problem of the evolution of altruism, with the payoffs now being units of fitness (often taken to be number of offspring). Natural selection favours individuals with a relatively high fitness compared with the rest of the population. Biological altruism, by definition, involves advantaging others at a

---

[3] In order to conclude that (*cooperate*, *cooperate*) is the best pair of strategies for them, the players have to judge the payoff combinations (*a*, *d*) and (*d*, *a*) to be worse 'for them' than (*b*, *b*).

|            |        | Player 2 |        |
|------------|--------|----------|--------|
|            |        | *Stag*   | *Rabbit* |
| Player 1   | *Stag* | s, s     | q, r   |
|            | *Rabbit* | r, q   | r, r   |

*s > r > q*

Figure 9.2. A stag hunt.

cost to self. So the altruist is at an evolutionary disadvantage compared with non-altruists and altruism should never evolve.

Arguably, many situations which have been analysed as prisoner's dilemmas can be better thought of as stag hunts.[4] Brian Skyrms (2004) considers the stag hunt, rather than the prisoner's dilemma, to be the paradigm problem of cooperation. The *stag hunt* is named for Rousseau's story about the beginnings of human society and the dilemma faced by hunters in the state of nature, as they begin to learn the benefits of cooperation. The generic payoff matrix for a stag hunt is given in Figure 9.2. For the story, consider two hunters who can each chase either stag or rabbit. Between the two of them, they could catch a stag which provides a large amount of meat, but if one of them chases stag alone then he will fail and go hungry. Rabbits can be caught by one person but provide less than half the meat of a stag. What each hunter cares about is how much meat he gets, so each prefers the outcome where both hunt stag to hunting rabbit (together or alone). In turn, hunting rabbit is preferred to hunting stag alone.

The game has two Nash equilibria, the outcome where both hunt stag and the outcome where both hunt rabbit. In other words, if Player 2 hunts stag then Player 1 does best if she also hunts stag and if Player 2 hunts rabbit then Player 1 does best if she hunts rabbit, and vice versa.

We might think of hunting stag as cooperating. Unlike the prisoner's dilemma, if one player cooperates then the other player is best off cooperating too. Both cooperating is an equilibrium. Further, hunting stag is better for both players than hunting rabbit so it may seem obvious that the players should coordinate on the equilibrium that they prefer.

However, from the assumptions that the players are perfectly rational (in the normal decision-theoretic sense of maximizing expected payoff)

---

[4] See Joshi et al. (2005) for an illustration from everyday life, Kollock (1998b) for an example from the lab, and Kollock (1998a) and Skyrms (2004) for theoretical arguments.

and that they have common knowledge of their rationality, we cannot deduce that each will choose *stag*. Or, expressing the same idea in normative terms, there is no sequence of steps of valid reasoning by which perfectly rational players can arrive at the conclusion that they ought to choose *stag*. This is because, from the assumption of rationality, all we can infer is that each player chooses the strategy that maximizes her expected payoff, given her beliefs about what the other player will do. All we can say in favour of *stag* is that, if either player expects the other to choose *stag*, then it is rational for the first player to choose *stag* too; thus, a shared expectation of *stag*-choosing is self-fulfilling among rational players. But exactly the same can be said about *rabbit*. And, in a one-shot game with common knowledge of rationality, there is no prior reason to expect the play of either *stag* or *rabbit*.[5]

This is an example of the problem of *equilibrium selection*. Classical game theory cannot make a unique prediction of play unless it is supplemented with new criteria to select between the equilibria. So even if the outcome where both play the cooperative strategy is a Nash equilibrium, orthodox game theory cannot predict its play – and sometimes it is not an equilibrium, as in the prisoner's dilemma. Analogously, although normative game theory assumes that if there is a rational way to play the game, then it will be a Nash equilibrium (Nash 1953), where there is more than one Nash equilibrium standard game theory has no way to advocate one over the other without introducing new assumptions – and, by this criterion, playing a strategy that is not a part of a Nash equilibrium is never rational.

## 9.2 TEAM REASONING AND COOPERATION

The prisoner's dilemma and the stag hunt are both games with scope. In the prisoner's dilemma, the Nash equilibrium is (*defect, defect*) but each player does better if the outcome is (*cooperate, cooperate*). In the stag hunt, (*rabbit, rabbit*) is a Nash equilibrium but the players do better under (*stag, stag*). The prisoner's dilemma seems more intractable because (*cooperate, cooperate*) is not a Nash equilibrium, but standard game theory has nothing to say about when or why hunters chase stag instead of rabbit without adding further equilibrium selection criteria. The equilibrium selection

[5] The stag hunt is sometimes called an 'assurance' game, as each player would play *stag* if assured that the other player were going to play stag too. It is rational for a player to play *stag* if she believes that the probability that the other player will play stag is greater than $(r - q)/(s - q)$, but the rational players of standard game theory have no reason to believe that.

problem is particularly perplexing in the stag hunt. Intuitively, it may seem obvious that each player should choose *stag* because both prefer the outcome of (*stag, stag*) to that of (*rabbit, rabbit*); but that 'because' has no standing in the formal theory.[6]

The source of both puzzles seems to be located in the mode of reasoning by which, in the standard theory, individuals move from preferences to decisions. In the syntax of game theory, each individual must ask separately 'What should *I* do?' In the stag hunt, the game-theoretic answer to this question is indeterminate. In the prisoner's dilemma, the answer is that *defect* should be chosen. Intuitively, however, it seems possible for the players to ask a different question: 'What should *we* do?' In the stag hunt, the answer to *this* question is surely: 'Choose (*stag, stag*)'. In the prisoner's dilemma, 'Choose (*cooperate, cooperate*)' seems to be at least credible as an answer.

The reasoning that occurs in game theory is *instrumental practical reasoning*, where conclusions about what an agent ought to do are inferred from premises that include propositions about what the agent is seeking to achieve. Such reasoning is *instrumental* in that it takes the standard of success as given; its conclusions are propositions about what the agent should do in order to be as successful as possible according to that standard.[7] So instrumental practical reasoning presupposes a unit of agency that pursues its own objectives. Standard game theory presumes that the unit of agency is the individual. Theories of team agency generalize game theory to allow the possibility that teams can be agents.[8]

The basic idea is that, when an individual reasons as a member of a team, she considers which *combination* of actions by members of the team would best promote the team's objective, and then performs her part of that combination. The rationality of each individual's action derives from the rationality of the joint action of the team. Gold and Sugden show that, in a group of agents, if there is common knowledge that each member 'group identifies', common knowledge that

[6] However, there is a counter-argument which suggests that rational hunters chase rabbit, formalized in Harsanyi and Selten's criterion of risk dominance (Harsanyi and Selten 1988). Intuitively, hunting rabbit means a player will never get the big prize but it also ensures that she will never go hungry.

[7] Bacharach, in his unpublished *Scientific Synopsis* describing his initial plans for *Beyond Individual Choice* (see Bacharach 2000), defines a mode of reasoning as valid in games if it is *success-promoting*: given any game of some very broad class, it yields only choices which tend to produce success, as measured by game payoffs.

[8] Individual reasoning is a special case of team reasoning, where the team has only one member. See the analysis in Gold and Sugden 2007a, 2007b.

each member aims to maximize the team payoff function and a unique profile of actions that does this, then each individual can reach the conclusion that she should choose her component of that profile (Gold and Sugden 2007a, 2007b).

This can lead to cooperation in the prisoner's dilemma and the stag hunt. First consider the prisoner's dilemma. We need to define a payoff function for the group consisting of Player 1 and Player 2. We shall assume that, when a player identifies with a group, she wants to promote the combined interests of its two members, at least in so far as those interests are affected by the game that is being played. If we assume that the payoff function treats the players symmetrically, we need to specify only three values of this function: the payoff when both players choose *cooperate*, which we denote $u_C$; the payoff when both choose *defect*, which we denote $u_D$; and the payoff when one chooses *cooperate* and one chooses *defect*, which we denote $u_F$ (for 'free riding'). It seems unexceptionable to assume that the team payoff function is increasing in individual payoffs, which implies $u_C > u_D$. Given the condition $b > (a + d)/2$, it is natural also to assume $u_C > u_F$. Then the profile of actions by Player 1 and Player 2 that uniquely maximizes the team payoff is (*cooperate, cooperate*). If there is common knowledge of the rules of the game, each player can use team reasoning to reach the conclusion that she should choose *cooperate* (Gold and Sugden 2007b).

Now consider the stag hunt. Again we need to specify only three values of this function: the payoff when both players choose *stag*, which I denote $u_S$; the payoff when both choose *rabbit*, which I denote $u_R$; and the payoff when one chooses *stag* and the other chooses *rabbit*, which I denote $u_{SR}$. It seems unexceptionable to assume that $U$ is increasing in individual payoffs, which implies $u_S > u_R > u_{SR}$. Then (*stag, stag*) is the profile that uniquely maximizes the payoff function, and so (provided there is common knowledge of the rules of the game), each player can use team reasoning to reach the conclusion that she should choose *stag*.

Thus team reasoning can predict the play of the the payoff dominant equilibrium. Payoff dominance is often used as a criterion for equilibrium selection (e.g. Harsanyi and Selten 1988). It is intuitively compelling but, previously, it has not been justified by standard game-theoretic rationality assumptions. Rather it is an additional supposition, included purely for the purposes of equilibrium selection. In contrast, team reasoning can explain why rational agents play the strategies that lead to the payoff dominant equilibrium (given a reasonable assumption about the team payoff function, discussed further below).

The idea that cooperating is better for the group also plays a fundamental role in the answer provided by theories of 'group' or multilevel selection to the evolutionary puzzle of altruism. These theories allow that natural selection can act on groups as well as individuals. Although altruism puts the individual altruist at an evolutionary disadvantage, it can be advantageous at the level of the group, putting the group at an advantage compared with groups composed of non-altruists. Depending on the relative strength of interindividual and intergroup evolutionary pressures, altruism may evolve because it is good for the group (Sober and Wilson 1998).

## 9.3 TEAM AGENCY, GROUP GOALS AND EXPECTED UTILITY THEORY

A minimal constraint on being an agent, for the purposes of instrumental practical reasoning, is that the (group) agent has an objective, which can be the basis of instrumental reasoning for its members. So theories of team reasoning need to assume that there is a group objective.[9] The basic idea of team reasoning places no constraints on this objective but, in order to operationalize the theory, it is necessary to make some assumptions about the group's goals. For example, in the discussion of cooperation above, I assumed that the collective utility function was increasing in individual payoffs, that it was symmetrical and, in the case of the prisoner's dilemma, I made an assumption about how it would rank outcomes with different distributions of individual utility.

A key issue is the relation between the individual agents' utility and the group's utility. Theories of team agency differ in the constraints they impose on this relationship. In this section, I compare what Bacharach and Sugden's theories advocate that players should do in the prisoner's dilemma when there is not common knowledge of group identification. I then use the comparison to correct two misconceptions about what the theory of team reasoning is committed to, as regards expected utility maximization and the group goal.

### 9.3.1 Team reasoning without common knowledge of group membership

Bacharach's theory of 'circumspect team reasoning' was formulated to explain cooperation in situations where there is not common knowledge

---

[9] Extant theories also assume that the group's objective is common knowledge among team reasoners, but that assumption could be relaxed.

of group identification. For Bacharach, whether a particular player identifies with a particular group is a matter of 'framing'. A *frame* is the set of concepts a player uses when thinking about her situation. In order to team reason, a player must have the concept 'we' in her frame. Bacharach proposes that the 'we' frame is normally induced or *primed* by games that have the property of *strong interdependence*. Roughly, a game has this property if it has a Nash equilibrium which is Pareto dominated by the outcome of some feasible strategy profile. So both the prisoner's dilemma and the stag hunt have the property of strong interdependence. (In the stag hunt, (*rabbit, rabbit*) is dominated by (*stag, stag*) – itself a Nash Equilibrium. In the prisoner's dilemma, (*defect, defect*) is Pareto dominated by (*cooperate, cooperate*).)

Although Bacharach proposes that the perception of strong interdependence increases the probability of group identification, he does not claim that games with this property *invariably* prime the 'we' frame. More specifically:

> In a Prisoner's Dilemma, players might see only, or most powerfully, the feature of common interest and reciprocal dependence which lie in the payoffs on the main diagonal. But they might see the problem in other ways. For example, someone might be struck by the thought that her coplayer is in a position to double-cross her by playing [*defect*] in the expectation that she will play [*cooperate*]. This perceived feature might inhibit group identification. (Bacharach 2006, p. 86)[10]

The implication is that the 'we' frame *might* be primed but, alternatively, a player may see the game as one to be played by two separate individual agents.

In Bacharach's theoretical framework, this dualism is best represented in terms of *circumspect team reasoning*. Suppose there is a random process which, independently for each member of the group, determines whether or not that individual identifies with the group. Let $\omega$ (where $0 < \omega \leq 1$) be the probability that, for any individual player, the 'we' frame comes to mind; if it does, the player identifies with the group. Then an individual who group identifies will maximize the *expected value* of the group payoff function given the probabilities that other group members fail to identify. (Any individual who identifies is assumed to also know the value of $\omega$.

---

[10] For Bacharach these features compete to be noticed and it is their relative salience that determines whether or not an agent will group identify. Another natural reading has the agent deliberating about whether or not to use the 'we' frame. As explained below, reasoning about frames has no place in Bacharach's framework. But Smerilli (2012) takes this 'double-crossing intuition' and proposes an extension to the theory, where agents adjudicate between the outcomes of individual and team reasoning, based on reasoning about deviation from equilibrium.

The idea is that, in coming to frame the situation as a problem 'for us', an individual also gains some sense of how likely it is that another individual would frame it in the same way; in this way, the value of $\omega$ becomes common knowledge among those who use this frame.)

We can apply this to the prisoner's dilemma played by the group of Player 1 and Player 2. Define the group payoff function as before. If the 'we' frame comes to mind, with probability $\omega$, the player identifies with the group. Assume that, if this frame does *not* come to mind, the player conceives of herself as a unit of agency and thus, using best-reply reasoning, chooses the dominant strategy *defect*. We can now ask which protocol maximizes the group payoff function, given the value of $\omega$. Viewed from within the 'we' frame, the protocol (*defect, defect*) gives a payoff of $u_D$ with certainty. Each of the protocols (*cooperate, defect*) and (*defect, cooperate*) gives an expected payoff of $\omega u_F + (1 - \omega)u_D$. The protocol (*cooperate, cooperate*) gives an expected payoff of $\omega^2 u_C + 2\omega(1 - \omega)u_F + (1 - \omega)^2 u_D$. There are two possible cases to consider. If $u_F \geq u_D$, then (*cooperate, cooperate*) is the team utility-maximizing protocol for all possible values of $\omega$. Alternatively, if $u_D > u_F$, which protocol maximizes the team utility function depends on the value of $\omega$. At high values of $\omega$, (*cooperate, cooperate*) is uniquely optimal; at low values, the uniquely optimal protocol is (*defect, defect*).[11]

If we assume *either* that $u_F \geq u_D$ or that the value of $\omega$ is high enough to make (*cooperate, cooperate*) the uniquely optimal protocol, we have a model in which players of the prisoner's dilemma choose *cooperate* if the 'we' frame comes to mind, and *defect* otherwise. Bacharach offers this result as an explanation of the observation that, in one-shot prisoner's dilemmas played under experimental conditions, each of *cooperate* and *defect* is usually chosen by a substantial proportion of players. He also sees it as consistent with the fact that there are many people who think it completely obvious that *cooperate* is the only rational choice, while there are also many who feel the same about *defect*.

If $\omega = 1$ then there is common knowledge of group identification among group members, if $\omega < 1$ then there is not. Since Bacharach's team reasoners act based on their *ex ante* probability that other members will group identify, if $\omega$ is less than 1 but high enough to make (*cooperate, cooperate*) the uniquely optimal protocol, a team reasoner may find *ex*

---

[11] We can normalize the payoff function by setting $u_C = 1$ and $u_D = 0$. Then, given that $u_F < 0$, the critical value of $\omega$ is $\omega^* = 2u_F/(2u_F - 1)$. The protocol (*cooperate, cooperate*) is optimal if and only if $\omega \geq \omega^*$, (*defect, defect*) is optimal if and only if $\omega \leq \omega^*$. There is no non-zero value of $\omega$ at which (*cooperate, defect*) or (*defect, cooperate*) is optimal.

*post* that she has cooperated when the other player has defected. In other words, she may find that she has been 'suckered' – as sometimes happens in the experiments whose results Bacharach sought to explain. This can happen because Bacharach's team reasoners cooperate without assurance that other group members are also team reasoning.[12]

Further, if $u_F \geq u_D$, then the team reasoner will cooperate regardless of whether she expects the other player to group identify, regarding it as better to be suckered by the other player than to defect. For Bacharach, the way the group utility function ranks the off-diagonal payoffs is an open question. In a very brief discussion of the group objective, Bacharach claims that it is likely to be Paretian and to embody principles of fairness (Bacharach 2006, p. 88). However these are not specified as conceptual constraints, but as testable hypotheses. Bacharach thought that group identification could explain why members of nationalistic movements are ready to sacrifice their lives for the cause (Bacharach 2006, p. 91, n. 4*), so he allowed in principle that the group objective might be welfare decreasing for some members.

Sugden disagrees with Bacharach on both these points. For Sugden, the purpose of the theory of team reasoning is to explain how people cooperate for mutual advantage, so he takes exception to the idea that the team utility function might make some individuals worse off by their individual lights (for example, ranking $u_F$ above $u_D$), and to the possibility that a team reasoner might cooperate without assurance that other group members will act likewise. For Sugden, a person should not be made worse off by team reasoning. Hence he would place more constraints on the group objective than Bacharach. In Sugden's theory the team should pursue outcomes that are advantageous to all its members.

It also follows that Sugden's team reasoners will not risk being suckered. On Sugden's account of *mutually assured team reasoning*, a person will not commit herself to team reasoning unless she has assurance that other team members will also act on team reasoning. Sugden uses a theoretical framework in which the central concept is *reason to believe*. To say that a person has reason to believe a proposition $p$ is to say that $p$ can be inferred from propositions that she accepts as true, using rules

---

[12] Bacharach also countenances a version of team reasoning that he calls *restricted team reasoning*, where the team reasoners know in advance that some of the team members will not team reason, and optimize the team utility function as best they can given that some team members will not function. But he prefers circumspect team reasoning because it is more general; we often don't know for certain whether or not other team members will group identify, or what those who do not group identify will do.

of inference that she accepts as valid. In mutually assured team reasoning, team members will not act on the results of team reasoning unless each has reason to believe of all the others that (1) they identify with the group and acknowledge the group payoff function as the objective of the group, and (2) they endorse and act on mutually assured team reasoning. So if Sugden's group members are not sure that they will all cooperate to achieve what they all take to be best for the group, then they will not team reason.

### 9.3.2 Team reasoning and expected utility theory

Some philosophers are uneasy about the association between team reasoning and expected utility theory. For instance, Raimo Tuomela (2009, p. 298), has complained of Gold and Sugden that 'the only collective goal that they consider and seem to take to be possible in their account is maximization of collective utility'. Hence Tuomela claims that team reasoning is not applicable to all cooperative contexts because its conception of a goal is not applicable. It is quite difficult to work out what Tuomela's complaint is, since there is a trivial sense in which virtually any reasoning can be represented as the maximization of a function so, when it occurs, team reasoning must involve maximization of a team objective function.[13] But his remarks reflect a tendency of some philosophers to wonder whether expected utility theory is the right framework to apply in all occurrences of team agency.

In response to that worry I note that, although Bacharach's theory of team reasoning explicitly incorporates expected utility theory, team reasoning per se is not committed to the idea of expected utility maximization. While Bacharach's team reasoners maximize a group objective given that some members do not group identify, Sugden's cooperate in a mutually advantageous enterprise. Hence Bacharach's theory implies that group members act to get the best outcome for the group even when other members fall short, whereas in Sugden's they only team reason when they are sure that their cooperative actions will be reciprocated. So, in the presence of uncertainty, Sugden's version of team reasoning does not involve maximization of expected utility.

---

[13] In making his inference Tuomela also implies that the conscious goal of the agents is to 'maximize utility'. But that would be a mistake because modelling a situation using game theory does not imply that agents have the conscious goal of maximizing utility. I address this mistake elsewhere (Gold, unpublished manuscript).

### 9.3.3 Team reasoning and averaging

There has been some presumption that team utility is the sum or average of the group utilities, the 'utilitarian' payoff function. Bacharach (1999), in an illustrative example of his theory, took the group payoff to be the mean of the individual payoffs, as does Alessandra Smerilli (2008). Colman et al. (2007), in their test of team reasoning, followed Bacharach's later suggestion that team reasoners will aim for outcomes that Pareto-dominate Nash equilibria, even when those outcomes are not equilibria themselves. Colman et al. point out that, as a consequence of this, the team reasoning outcomes in their experiment maximize the sum of the payoffs of the players.

However, it should be obvious from the discussion above that team reasoning does not require that the team utility is the sum or average of the utility of the individual members and that some specific versions may reject this constraint.

As we saw, Bacharach hypothesized that the team utility function would be Paretian, i.e. if every individual agent gets at least as much utility in outcome $x$ as outcome $y$, and at least one agent does strictly better, then the group function will rank outcome $x$ above outcome $y$. The utilitarian payoff function is Paretian but so are many others. The Pareto criterion alone cannot resolve the question of how the team function ranks outcomes in situations of partial, but not complete, harmony of interest – for instance it does not give any guidance about how to rank the off-diagonal payoffs in the prisoner's dilemma. In contrast the utilitarian function can provide a ranking. However, the utilitarian function is more informationally demanding than the Pareto criterion. Applying the Pareto criterion only requires that utility is ordinally measurable and does not require interpersonal comparability, whereas the utilitarian function requires interpersonal comparability of utility.

Although Bacharach posited the Pareto criterion as an empirical hypothesis, he did not specify any conceptual constraints on the group goal. In contrast, Sugden would impose conceptual constraints on the team utility function which lead to the rejection of the idea that it is the average of the utility of the team members. Sugden assumes that, when a player identifies with a group, she wants to promote the combined interests of its two members, at least in so far as those interests are affected by the game that is being played. Since his theory is concerned with cooperation for mutual advantage, team members will not take on a team utility function that would make them worse off than they would be as individual agents (cf. the discussion in Section 9.2 of the team payoff function in the prisoner's dilemma). Since a team utility function that maximizes the sum of the individual payoffs takes no account of the distribution of payoffs between players, or of whether individual players improve their lot by team reasoning, the team utility function in Sugden's theory cannot simply be the sum of individual payoffs.

There is an obvious analogy between the notion of a team payoff in the theory of team reasoning and the idea of group fitness in evolutionary theories of group selection.[14] The issue of the relation between individual and team payoff also has an interesting analogy with evolutionary biology. In evolutionary theory, most formal models define group fitness as total or average individual fitness. However, Rick Michod (2005) has argued that the fitness of the group cannot be equated with the total fitness of its parts. Michod allows that one situation can have more group fitness than another, even when the fitness of the individual units is exactly the same in each. Hence his model violates the Pareto criterion of social choice theory (Okasha 2009) – the one criterion that it seems advocates of team reasoning agree on!

## 9.4 FRAMING AND TEAM REASONING

Bacharach and Sugden's theories also differ in the way that group agency comes about. Gold and Sugden (2007a, 2007b) labelled Bacharach's theory 'team agency as the result of framing', in contrast to Sugden's emphasis on assurance. But that should not be taken to imply that framing has no role to play in Sugden's theory. By comparing how agents come to team reason in Bacharach and Sugden's theories, I will unpick the various processes involved in framing in decision making in general.

Although Bacharach's theory is most strongly associated with framing, there is a framing step in Sugden's theory as well. Sugden (2000) makes an analogy between the way that, in the theory of team reasoning, people have different preferences from the perspectives of different units of agency and the way that, in standard decision theory, preferences are relative to particular conceptions of, or framings of, decision problems. He also explains group identification in terms of framing:

---

[14] This was suggested to me by Samir Okasha, who himself draws an analogy between individual and group fitness, and individual and social preference orderings as investigated in social choice theory (Okasha 2009).

The idea is that, in relation to a specific decision problem, an individual may conceive of herself as a member of a group or team, and conceive of the decision problem, not as a problem for her but as a problem for the team. In other words, the individual frames the problem, not as 'What should I do?', but as 'What should we do?' (2000, pp. 182–183)

Sugden describes the framing of a decision problem as reflecting the agent's subjective perceptions – what she takes her decision problem to be – and her preferences as her all-things-considered choice-relevant reasons. He says that her preferences are defined relative to her framing of the problem. So when Sugden says that an agent frames the problem as a problem for the team, that implies that the agent sees team considerations and achieving the group goal as choice relevant. Hence Sugden's account of team reasoning also involves framing, in the sense that it presupposes that the potential team reasoner conceptualizes the situation as a problem for the team, i.e. sees the team utility function as representing choice-relevant reasons. However, this is not enough to ensure that the agent will team reason. The agent also has to decide that she endorses team reasoning.

Sugden likens endorsing mutually assured team reasoning to making a unilateral commitment to a certain form of practical reasoning, where this reasoning does not generate any implications for action unless one has assurance that others have made the same commitment. Such assurance could be created by public acts of commitment or induced by repeated experience of regularities of behaviour in a population. But the questions of assurance and endorsement are separate: even if each individual were assured that others would choose their components of the group payoff-maximizing profile, each would still have to decide whether team reasoning was a mode of reasoning that she wanted to endorse. It is possible for a person either to have assurance but not endorse team reasoning, or to endorse team reasoning but not to have assurance. Assurance seems to be bound up with group identifying: 'to construe oneself as a member of a team, one must have some confidence that the other members of that team construe themselves as members too' (Sugden 2000, p. 194). This suggests that an agent could see the relevance of team-directed reasons but, because she does not have assurance that other agents will team reason, she does not group identify, in the sense of conceiving the group as a unit of agency, acting as a single entity in pursuit of some single objective (Gold and Sugden 2007a, 2007b).

This contrasts with Bacharach's approach in *Beyond Individual Choice*, where he is interested in 'the role of spontaneous group identification in decision making' (Bacharach 2006, p. 81), and there is no gap between framing the decision problem as a problem for 'us' and group identifying, or between group identifying and team reasoning, in which the agent can choose whether to group identify or to team reason.[15] Bacharach does recognize that these are simplifications. When mooting the interpretation of $\omega$ as the probability that someone group identifies he footnotes an unexplored subtlety, that '[t]his assumes that group identification, if it happens, primes [team reasoning] with probability 1. More generally, if this probability is $p < 1$, and that of group identification is $v$; then $\omega = pv$.' (2006, p. 152, n. 14). Nevertheless, reasoning about whether or not to team reason does not enter the picture.

Nor does an agent who notices 'we' reasons get to decide whether or not she identifies with the group. When talking about group identification, Bacharach makes a distinction between the 'salience' of features that tend to promote group identity and their 'effectiveness', i.e. their tendencies, if and when perceived, to stimulate or inhibit group identity. However, Bacharach speculates that there may be 'a positive relationship between salience and effectiveness', that if features tending to promote group identity are highly salient then, when noticed, they are also highly effective (Bacharach 2006, p. 87). So there is conceptual space in Bacharach's theory for a gap between noticing the group and group identifying. Elsewhere Bacharach says that this gap is filled by *affiliation*, 'a psychological process in which a person who does think about a certain group, defined by some shared property, comes to think about it as "us"' (Bacharach 1997, p. 2). Thus the gap is filled by a psychological process, not a choice.

For both Bacharach and Sugden there are potential gaps between noticing the group and group identifying, and between group identifying and team reasoning. For Sugden these gaps are both bridged by decisions, for Bacharach they are filled by psychological processes. As explained below, for neither of them are these points subject to standards of practical, instrumental rationality.

Now we are in a position to delineate the various steps involved in framing and decision making. In the context of team reasoning, first the agent must see the possibility of cooperation and notice the potential for team reasoning, then she must group identify and see the group goal as providing a choice-relevant reason; third, she must decide to act on team

---

[15] Bacharach set out an earlier version of the theory in a 1997 working paper. Discussion of differences between the two will be confined to the footnotes.

reasoning. The same steps occur in framing in general. We do not notice all the ways that we, counterfactually, could distinguish between objects or between actions. For a distinction to be the basis of action, an agent must first notice it. The move from 'noticing' *simpliciter* to 'noticing as choice relevant' corresponds to Bacharach's 'effectiveness'. It also takes us to Sugden's start point, where an agent conceptualizes her decision problem using considerations that she takes to be choice relevant. The move from 'noticing as choice relevant' to deciding to act on team reasons corresponds to Sugden's 'endorsement'. We might call the two steps together, that take the agent from noticing a feature/concept cluster/family to acting on the reasons that it provides, 'motivational grip'. Then the two steps provide two reasons why a concept cluster, even when noticed, would not have motivational grip and hence would not be the basis for action: because it is not perceived as choice relevant or because it is seen as choice relevant but, nonetheless, does not form the basis of the agent's action. One reason an agent might not act on a feature that she acknowledges as choice relevant is that she decides that it is not a reason that she wants to endorse. Other reasons why she might fail to move from seeing as choice relevant to acting include making mistakes and akrasia (weakness of will).

These steps – noticing, noticing as choice relevant, and acting on – are involved in framing more generally, not just in the framing that leads to team reasoning.[16] To illustrate this with an example, inspired by Sugden's discussion of how economists model demand for consumption goods (Sugden 2000), consider someone in a supermarket deciding whether to buy an apple or an orange. It is natural to think of her as conceptualizing her decision problem as choosing whether to 'buy an apple' or 'buy an orange'. In fact, she is probably confronted by a crate of apples and a crate of oranges, each containing numerous pieces of fruit. But, although we can distinguish many pieces of fruit according to their position, a normal person wouldn't see her choice as being between 'buy the apple on the top left', 'buy the apple, top left but one', ... 'buy the orange on the bottom right but one', 'buy the orange on the bottom right'. The position of the pieces of fruit is not relevant to the decision problem and is not a part of its conceptualization. But position is a feature that someone might notice and discard as irrelevant to the choice.

---

[16] I rely here on an intuitive grasp of the notion of 'noticing' but I note that it is complicated and there are conceptual issues regarding noticing that I do not have space to go into here (e.g. we tend to notice things that are choice-relevant).

As Sugden points out, questions about how to frame decision problems do not always have easy answers. Another way that we can distinguish fruit in supermarkets is by the identity of its supplier. Like position, the supplier does not usually enter models of purchasing decisions. However, in the 1980s whether the orange came from South Africa was, for many people, choice-relevant information. Nowadays some people choose e.g. to buy products with fewer food miles or to buy products that are fair trade. But other consumers are not bothered by the place of origin of their food. Supermarkets often label where food is from but many people do not assimilate this information, they do not 'notice' it. In order for place of origin to influence purchasing decision, a shopper must first notice the origin, then take place of origin to be a choice-relevant feature and, finally, act on that feature. Further, people who are already committed to not buying from certain suppliers are more likely to notice and even seek out information about the place of origin of the goods in their shopping basket.

Should consumers discriminate oranges according to their suppliers? For Sugden (2000, p. 200), 'there is no objective answer to this question, independent of consumers' subjective perceptions: what matters is whether consumers take them to be the same'. This reflects Sugden's general approach, which does not acknowledge agent-neutral concepts of 'validity' and 'rationality'. I won't comment upon his approach here. I will merely note that, in saying that there are no objective answers to questions about framing, Sugden places himself at odds with most of the literature on framing in individual decision making which, at least implicitly – and sometimes explicitly – assumes that there is a correct way to frame a decision problem. I also note that, despite many researchers' implicit commitments, there is at present no theory of rational framing that tells us whether a particular framing of a decision problem is 'correct' or not.[17]

In *Beyond Individual Choice* Bacharach does not admit questions about the rationality of frames either. He is investigating valid instrumental practical reasoning. The agent can only reason from premises about the world that are accessible to her, so frames set the parameters of reasoning. Frames are supplied by involuntary psychological processes, to which the

---

[17] James Joyce (1999) makes some intriguing comments about how such a theory might proceed. John Broome (1991) has argued that we should individuate outcomes by justifiers, where a *justifier* is a difference between two putative outcomes that makes it rational to have a preference between them. This implies that framing is amenable to rational assessment, but Broome does not have any further discussion of what differences it is rational to have a preference between.

concept of practical rationality does not apply. In earlier papers, Bacharach suggested that a complete theory of rational choice would address questions about whether particular framings are rational or not (Bacharach 2000) and discussed various problem cases (Bacharach 1998). But any principles of rationality that are concerned with how agents should frame the world will not be principles of instrumental rationality. John Broome's rational requirements of indifference (Broome 1991), Susan Hurley's agent neutral goals (Hurley 1989) and Elizabeth Anderson's principles of rational self-identification (Anderson 2001) all rely on thicker notions of rationality than means–end reasoning.

## 9.5 TEAM REASONING VS OTHER THEORIES OF COOPERATION

Team reasoning has been advanced as a theory that can explain why cooperation is rational, as a theory that can predict cooperation, and as a psychological theory about how people make decisions. In this section I will compare team reasoning with other theories of cooperation on these dimensions. Again, the starting point for this is a comparison of the theories' analyses of cooperation in the prisoner's dilemma.

### 9.5.1 *The prediction of rational cooperation*

In the analysis above, a rational player of the one-shot prisoner's dilemma can choose *cooperate*. For many game theorists, this conclusion is close to heresy. For example, Ken Binmore (1994, pp. 102–117, quotation from p. 114) argues that it can be reached only by 'a wrong analysis of the wrong game': if two players truly face the game shown in Figure 9.1, then it follows from the meaning of 'payoff' and from an unexceptionable concept of rationality that a rational player must choose *defect*. Binmore recognizes that rational individuals may sometimes choose *cooperate* in games in which *material payoffs* – that is, outcomes described in terms of units of commodities which people normally prefer to have more of rather than less, such as money, or years of not being in prison – are as in Figure 9.1. But that just shows that the payoffs that are relevant for game theory – the payoffs that govern behaviour – differ from the material ones. The first stage in a game-theoretic analysis of a real-life situation should be to find a formal game that correctly represents that situation.

Thus in response to the problem of explaining why *cooperate* is sometimes chosen in games whose material payoffs have the prisoner's dilemma structure, the methodological strategy advocated by Binmore is that of *payoff transformation*: we should look for some way of transforming material payoffs into game-theoretic ones that makes observed behaviour consistent with conventional game-theoretic analysis. This strategy has been followed by various theorists who have proposed transformations of material payoffs to take account of psychological or moral motivations that go beyond simple self-interest. For example, Ernst Fehr and Klaus Schmidt (1999) and Gary Bolton and Axel Ockenfels (2000) have proposed that, for any given level of material payoff for any individual, that individual dislikes being either better off or worse off than other people. Matthew Rabin (1993) proposes that each individual likes to benefit people who act with the intention of benefiting him, and likes to harm people who act with the intention of harming him. Cristina Bicchieri (2006) proposes a theory of norms such that, when an individual recognizes that she is in a situation where a norm exists, she prefers to conform to the norm if she believes that a sufficiently large number of other people will also conform and that a sufficiently large number expect her to conform.

The theory of team reasoning can accept Binmore's instrumental conception of rationality, but rejects his implicit assumption that agency is necessarily vested in individuals. We can interpret the payoffs of a game as showing what each player wants to achieve *if she takes herself to be an individual agent*. In this sense, the interpretation of the payoffs is similar to that used by Binmore: payoffs are defined, not in material terms, but in terms of what individuals are seeking to achieve. The theory of team reasoning can replicate Binmore's analysis when it is applied to players who take themselves to be individual agents: if Player 1 frames the game as a problem 'for me', the only rational choice is *defect*. However, the theory also allows the possibility that Player 1 frames the game as a problem 'for us'. In this case, the payoffs that are relevant in determining what it is rational for Player 1 to do are measures of what she wants to achieve *as a member of the group {Player 1, Player 2}*; and these need not be the same as the payoffs in the standard description of the game.

Thus, there is a sense in which team reasoning, as an explanation of the choice of *cooperate* in the prisoner's dilemma, depends on a transformation of payoffs from those shown in Figure 9.1. However, the kind of transformation used by theories of team reasoning is quite different from that used by theorists such as Fehr and Schmidt. In team reasoning, the transformation is not from material payoffs to choice-governing payoffs; it is from payoffs which govern choices for one unit of agency to payoffs which govern choices for another. Thus, payoff transformation takes place as part of a more fundamental *agency transformation*. Once we accept the possibility of agency transformation, in a situation where there is common knowledge of group identification all that is needed to make a unique prediction of rational cooperation is the assumption that (*cooperate, cooperate*) is the best profile of actions for the two players together. That assumption is hardly controversial in the games under consideration and in many everyday situations.[18]

Then, the idea is that the rationality of each individual's action derives from the rationality of the joint action of the team. Bacharach says,

> The remainder of the team reasoning procedure is then inevitable. Once I have computed the best team profile and identified my component in it, team reasoning prescribes that I should choose to perform this component ... I am rationally obliged to follow the remainder of the procedure. If I believe that *we* should do a certain combination of actions, it is logically required that I also believe that I should do the bit that falls to me. If I am convinced that we should pass each other on the left, I must also think that I should pass you on the left (and that you should do likewise). The underlying general principle is that I cannot coherently will something without willing what I know to be logically entailed by it. This is a standard inference rule of deontic logic, the logic of what ought to be. (2006, p. 136)[19]

Not everyone agrees with Bacharach about the rules of deontic logic. But even those who disagree that it follows from the rationality of a profile *for the group* that it is rational for the *individual members* to play their part must acknowledge that team reasoning is no worse off than standard decision theory in this respect. Actions are taken by temporal parts of people and, as is notorious in the theory of dynamic choice, the incentives of a temporal part may differ from the interests of her later selves or from the interests of the agent conceived as a person over time. Any difficulty that the theory of team reasoning has when explaining why it is rational

---

[18] There is also a sense in which team reasoning is more efficient than payoff transformation: it is more efficient at maximizing the team payoff function than a game between two *benefactors* (individuals who have each taken on the team payoff function but still use individual reasoning). Bacharach (1999) proves that optimal team decision rules give Nash equilibra for benefactors but not all Nash equilibria for benefactors define optimal team decision rules. Intuitively, team reasoners are concerned with co-ordination, and co-ordination is important for efficiency – there is usually a lesser total payoff when agents fail to co-ordinate in games with scope. The team-reasoning equilibrium singles out the best of the possible ways of co-ordinating, whereas benefaction generally does not lead to a unique equilibrium, leaving the possibility of miscoordination.

[19] Anderson also argues that, when a group of agents see their actions as jointly advancing a common goal, then it is rational for an agent 'to do my part in what we are willing together' (Anderson 2001, pp. 28–30).

for individuals to play their part in the team plan is analogous to the difficulty that standard decision theory has to overcome, if it is to explain why it is rational for temporal parts to act in the interests of the person over time.[20]

In contrast, even if payoffs have been transformed to go beyond self-interest, conventional game theory still does not necessarily provide an explanation of why each individual chooses *cooperate* in a prisoner's dilemma. The payoff transformation theories listed above all tend to function by decreasing the relative appeal of the 'off-diagonal outcomes' where one player cooperates and the other doesn't. Depending on the theory, the individual does not want: to get a different payoff from the other player; to cooperate and hence benefit another player who is defecting and hence harming himself, and vice versa; or to follow a norm when no-one else does. This can make (*cooperate, cooperate*) into a Nash equilibrium. But it does not tend to alter the equilibrium status of (*defect, defect*).[21] There are now two Nash equilibria, (*cooperate, cooperate*) and (*defect, defect*). So payoff transformation theories face the problem of equilibrium selection, explained above (Sections 9.1–2). They cannot show that rational players of this game choose *cooperate* or make a unique prediction of cooperation. The situation of payoff transformation theories is similar regarding cooperation in the stag hunt, except that (*stag, stag*) was already an equilibrium anyway, before any payoff transformation occurred.[22]

The problem faced by theories of individual reasoning in explaining cooperation is that recommendations for action are conditional on the actor's beliefs about what the other individuals will do. Within classical theory, there are no grounds for assigning such beliefs. However payoff transformation theories can also predict cooperation *if* the prior probability that each player assigns to the other cooperating is suitably high or, for Bicchieri, if the probability that the other player is sensitive to the norm is suitably high. So, in order to make the prediction, they must be supplemented with an account of the probabilities that the agent assigns. The agents of classical game theory have no grounds for assigning

---

[20] For more on the analogy between individuals and teams, and time slices and persons-over-time, see Gold (2012).

[21] For a more detailed exposition of team reasoning versus payoff transformation theories of cooperation in the prisoner's dilemma, including a numerical example, see section 3 of Gold and Sugden (2007b).

[22] It is possible that, in the stag hunt, a player who was suckered when playing stag would not consider that the rabbit player *intended* to harm her. However, even if Rabin's theory does not downgrade the off-diagonal outcomes in this case, it is not going to show that (*stag, stag*) is uniquely rational, because the theory does not change the equilibrium status of (*rabbit, rabbit*).

probabilities one way or the other so, if the proponents of payoff transformation theories wish to claim that they predict rational cooperation,
then they also owe us an account of how rational agents acquire their
probability estimates. It is usually claimed that rational probabilities are
formed by Bayesian updating, and Bicchieri explicitly appeals to the idea
that the probability that the other player will be norm-sensitive is high
in a Bayesian game.[23] In contrast, the theory of team reasoning generates
recommendations for action that are not conditional on the actor's beliefs
about what the other individuals will do. Where there is common knowledge of group identification, the theory of team reasoning can predict
rational cooperation. Recommendations are conditional on beliefs about
group identification, but these are outside the purview of instrumental
rationality. While payoff transformation theories of cooperation can predict cooperation given suitable beliefs, there is a sense in which only team
reasoning predicts rational cooperation, within the restrictions of classical
game theory.

### 9.5.2 The explanation of cooperation

Team reasoning is proposed by both Bacharach and Sugden as a mode
of reasoning that people actually use. Advocates of payoff transformation
theories also claim that their favoured payoff transformation describes
what people do. All these theories can, in principle, explain cooperation.
But which provides the best explanation of what people do? One way to
adjudicate the question is to compare the theories with respect to evidence about their auxiliary hypotheses.[24]

One type of auxiliary hypothesis relates to evolution. Any proposed
proximate mechanism for cooperation must be a possible result of evolution, so having a plausible evolutionary story speaks in favour of any
particular proximate mechanism. (Conversely, if we have independent

---

[23] While considering Bayes and rationality, we might also note that framing is problematic for
Bayesianism. For the operation of Bayes rule, we must assume that the agent assigns a strictly
positive probability to each alternative. There is no space for an agent to be simply unaware of an
alternative.

[24] Another way is to run experiments to test the theories. However, since all the theories claim
to explain cooperation, when trying to discriminate between them, looking at behaviour in
standard games is not sufficient. Some effort has been put into identifying which reasoning or
payoff transformations people actually use and some ingenious experiments have been devised.
Colman et al. (2007) test team reasoning by constructing games in which there are unique Nash
equilibria that would not maximize a team payoff function. Guala et al. (2009) compare team
reasoning with other theories of cooperation by considering the role of expectations in various
theories and manipulating subjects' beliefs.

reasons for favouring a particular proximate mechanism and if that mechanism has implications for the correct ultimate mechanism, then we might be able to use what we know about the proximate mechanism to adjudicate between ultimate, evolutionary explanations.) There is not space here to go into detail about the relation between proximate and ultimate explanations, but I will compare Bacharach and Sugden's views about the evolution of the capacity for team reasoning.

Bacharach (2006, ch. 3) argued that group selection can explain cooperative behaviour and that group identification is a key proximate mechanism in producing well-functioning groups. Sugden, on the other hand, thinks that group selection is unlikely to be part of any ultimate explanation, because the conditions that are required for group selection to occur were never fulfilled (R. Sugden, personal communication).

Sugden (2002, 2005) has suggested that team reasoning is related to generalized reciprocity, which is at least partly sustained by agents taking pleasure in the correspondence of their sentiments with the sentiments of others in the group, or *fellow-feeling*. A natural corollary would be to hypothesize that reciprocity is the appropriate ultimate explanation. Alternatively, Sugden's account of fellow-feeling might be compatible with kin selection. We might have evolved fellow-feeling because it was helpful for increasing the number of descendants left by our kin – who in Pleistocene times would also have been our fellow group (or *deme*) members. But, in modern times, the trait 'misfires' and we apply it to group members who are not related to us.

Alternatively, given the right environmental conditions, team reasoning might have evolved by individual selection. Although he mainly focuses on group selection, Bacharach himself raises this intriguing possibility. He claims that our ancestors faced *ludic diversity*, or an environment in which they had to play many types of games. Team reasoning gets good results in a variety of games and, since we have limited cognitive resources, team reasoners might have had an evolutionary advantage because team reasoning is a parsimonious way of achieving good outcomes across a diverse range of situations. In this story, team reasoning emerges by individual selection.

Hence it should be obvious that it is the case neither that group selection being wrong would be problematic for Bacharach's theory, nor that group selection being right would be problematic for Sugden. When Bacharach proposed that team reasoning evolved by group selection, he argued that a 'how possible' evolutionary explanation provided evidential support for the existence of team reasoning. The fact that there are multiple 'how possible' explanations available that support the evolution of team reasoning is not damaging to the theory. Although Bacharach and Sugden may prefer different 'how possible' explanations, it is not clear that either of their theories requires their preferred evolutionary explanation. In particular, whether or not team reasoning is a mechanism that people use is independent of whether or not there was group selection.

A second type of auxiliary hypothesis is about the psychological capacities that each theory requires. For example, sympathy is required for altruism, intention detecting for Rabin's kindness, a sense of obligation or desire for conformity for norms, group identification or fellow-feeling for team reasoning. Evidence for all of these capacities can be found. This supports the intuitively appealing view that humans are creatures with heterogenous motivations, so in practice a catholic approach is to be favoured. Gold and Sugden (2007a) explicitly say that a given pattern of behaviour could be generated either collectively, using team reasoning, or individually, using individual reasoning supplemented with appropriate beliefs.[25]

However, if we want to encourage cooperation then we need more precise answers, as different theories indicate different strategies for increasing cooperation: should we institute policies that affect expectations, increase group identification, or enable transparent perception of people's intentions? Even accepting that all the above motivations are used in some circumstances, a complete theory would specify which motivation is used in which circumstances.

Another type of auxiliary hypothesis that any of these theories could be supplemented with is a hypothesis about framing. In any particular situation that offers the potential for cooperation, it might be possible to construe the situation in several ways, so that more than one of altruism, inequality aversion, kindness, norms, team reasoning, etc., could be seen as choice relevant. As discussed above, a prerequisite for modelling a situation is deciding which considerations are choice relevant for the agent. Those theories that do not explicitly include framing assume it implicitly. If we accept that humans have heterogeneous motivations, then we need to know under what circumstances people frame the situation so that

---

[25] Another type of auxiliary hypothesis involves support for the cognitive primitives implied by a theory. With respect to team reasoning, Gold and Harbour (2012) argue that the primitives needed for team reasoning enjoy direct and diverse support from linguistic theory. They use this approach to discriminate between theories of collective intentions but, at present, this approach has not been used to discriminate between theories of cooperation.

each potential motivation is seen as being choice relevant. A general theory of framing might give us a theoretical grip on when each theory can predict choice, and provide testable hypotheses.

## 9.6 CONCLUSION

I have explained why games with scope pose both a normative and an explanatory problem for classical game theory, and I have shown how team reasoning can explain cooperation in prisoner's dilemmas and stag hunts. By comparing Bacharach and Sugden's theories of team reasoning, I have shown some of the ways in which theories of team reasoning can differ.

I clarified that neither Bacharach nor Sugden is committed to the idea that the team payoff function is the average of the individual utilities, and that Sugden is not committed to expected utility maximization. In doing so, I drew a structural analogy between models of team reasoning and models of group selection, although I demonstrated that the truth of the theory of team reasoning does not rely on the truth of the theory of group selection.

I also showed how framing has a role to play in both Bacharach's and Sugden's theories and, indeed, in all theories of cooperative motivations. A complete theory of cooperative behaviour would require an account of when people 'notice' particular features of situations, when people take features as 'choice relevant' and when noticing a feature as choice relevant leads to choosing in accordance with it.

## REFERENCES

Anderson, E. (2001). Unstrapping the straitjacket of 'preference': a comment on Amartya Sen's contributions to philosophy and economics. *Economics and Philosophy* 17, 21–38.

Bacharach, M. (1997). 'We' equilibria: a variable frame theory of cooperation. Working paper, Institute of Economics and Statistics, University of Oxford.

(1998). Preferenze razionali e descrizioni. In M. Galarotti and G. Gambetta (eds.), *Epistemologia et Economia*. Bologna: CLUEB.

(1999). Interactive team reasoning: a contribution to the theory of cooperation. *Research in Economics* 53, 117–147.

(2000). Scientific synopsis. Unpublished manuscript (describing initial plans for *Beyond Individual Choice* (Bacharach 2006)).

(2006). *Beyond Individual Choice: Teams and frames in game theory*. Princeton University Press.

Bardsley, N. (2007). On collective intentions: collective action in economics and philosophy. *Synthese* 157(2), 18.

Bicchieri, C. (2006). *The Grammar of Society: The Nature and Dynamics of Social Norms*. Cambridge: Cambridge University Press.

Binmore, K. (1994). *Game Theory and the Social Contract*, volume 1. Cambridge MA: MIT Press.

Bolton, G. and Ockenfels, A. (2000). ERC – a theory of equity, reciprocity and competition. *American Economic Review* 90, 166–193.

Broome, J. (1991). *Weighing Goods*. Oxford: Basil Blackwell.

Colman, A., Pulford, B. and Rose, J. (2007). Collective rationality in interactive decisions: evidence for team reasoning. *Acta Psychologica* 128, 387–397.

Fehr, E., and Schmidt, K. (1999). A theory of fairness, competition and cooperation. *Quarterly Journal of Economics* 114, 817–868.

Gilbert, M. P. (1987). *On Social Facts*. New York: Routledge.

Gold, N. (forthcoming). Framing and self-control. In N. Levy (ed.), *Self-Control and Addiction: Lessons from Philosophy and Cognitive Science*. Oxford University Press.

(unpublished manuscript). Group goals, game theoretic reasoning and spontaneous collective intentions.

Gold, N. and Harbour, D. (2012). Cognitive primitives of collective intentions: linguistic evidence of our mental ontology. *Mind & Language* 27(2), 109–134.

Gold, N. and Sugden, R. (2007a). Collective intentions and team agency. *Journal of Philosophy* 104(3), 109–137.

(2007b). Theories of team agency. In F. Peter and S. Schmidt (eds.), *Rationality and Commitment*. Oxford University Press.

Guala, F., Mittone, L., and Ploner, M. (2009). Group membership, team preferences, and expectations. CEEL Working Papers 0906, Cognitive and Experimental Economics Laboratory, Department of Economics, University of Trento, Italia.

Harsanyi, J., and Selten, R. (1988). *A General Theory of Equilibrium Selection in Games*. Cambridge, MA: MIT Press.

Hodgson, D. (1967). *Consequences of Utilitarianism*. New York: Oxford University Press.

Hollis, M. (1998). *Trust Within Reason*. Cambridge University Press.

Hurley, S. (1989). *Natural Reasons*. New York: Oxford University Press.

Joshi, M. S., Joshi, V., and Lamb, R. (2005). The prisoners' dilemma and city-centre traffic. *Oxford Economic Papers* 57(1), 70–89.

Joyce, J. (1999). *The Foundations of Causal Decision Theory*. Cambridge University Press.

Kollock, P. (1998a). Social dilemmas: the anatomy of cooperation. *Annual Review of Sociology* 24, 183–214.

(1998b). Transforming social dilemmas: group identity and cooperation. In P. A. Danielson (ed.), *Modeling Rationality, Morality and Evolution* (pp. 186–210). Oxford University Press.

Michod, R. (2005). The group covariance effect and fitness trade-offs during evolutionary transitions in individuality. *Proceedings of the National Academy of Science* 103(24), 9113–9117.

Nash, J. (1953). Two-person cooperative games. *Econometrica* 21, 128–140.

Okasha, S. (2009). Individuals, groups, fitness and utility: multi-level selection meets social choice theory. *Biology and Philosophy* 24, 561–584.

Rabin, M. (1993). Incorporating fairness into game theory and economics. *American Economic Review* 83, 1281–1302.

Regan, D. (1980). *Utilitarianism and Cooperation.* New York: Oxford University Press.

Skyrms, B. (2004). *The Stag Hunt and Evolution of Social Structure.* Cambridge University Press.

Smerilli, A. (2012). We-thinking and 'double-crossing': frames, reasoning and equilibria. *Theory and Decision*, www.springerlink.com/content/3347v4163114hou2.

Sober, E., and Wilson, D. S. (1998). *Unto Others: The Evolution and Psychology of Unselfish Behavior.* Cambridge, MA: Harvard University Press.

Sugden, R. (1993). Thinking as a team: toward an explanation of nonselfish behavior. *Social Philosophy and Policy* 10, 69–89.

(2000). Team preferences. *Economics and Philosophy* 16, 175–204.

(2002). Beyond sympathy and empathy: Adam Smith's concept of fellow-feeling. *Economics and Philosophy* 18, 63–87.

(2003). The logic of team reasoning. *Philosophical Explorations* 6, 165–181.

(2005). Fellow-feeling. In B. Gui and R. Sugden (eds.), *Economics and Social Interaction* (pp. 52–75). Cambridge University Press.

Taylor, M. (1987). *The Possibility of Cooperation.* Cambridge: Cambridge University Press.

Tuomela, R. (2009). Collective intentions and game theory. *Journal of Philosophy* 106, 292–300.

# CHAPTER 10

# An evolutionary perspective on the unification of the behavioral sciences

*Herbert Gintis*

The behavioral sciences have distinct research foci, but they include four conflicting models of decision making. The four are the economic, the sociological, the biological, and the psychological. These four models are not only different, which is to be expected given their distinct explanatory goals, but *incompatible*. In fact all four are flawed, but can be modified to produce a common framework for modeling choice and strategic interaction.

The economic model is rational choice theory, which takes the individual as maximizing a self-regarding preference-ordering subject to an unanalyzed and pregiven set of beliefs, or subjective priors. The sociological model is that of the pliant individual who internalizes the norms and values of society and chooses according to the dictates of the social roles he occupies. The biological model is that of the fitness maximizer who is the product of a long process of Darwinian evolution. The psychological model is that of the irrational and incompetent decision maker, incapable of applying Bayesian reasoning and ignorant of the rules of basic logical inference.

My framework for unification includes five conceptual units: (a) gene–culture coevolution; (b) the sociopsychological theory of norms; (c) game theory; (d) the rational actor model; and (e) complexity theory. Gene–culture coevolution comes from the biological theory of social organization (sociobiology), and is foundational because *Homo sapiens* is an evolved, biological, highly social, species. The sociopsychological theory of norms includes fundamental insights from sociology that apply to all forms of human social organization, from hunter–gatherer to advanced technological societies. These societies are the product of gene–culture coevolution, but have emergent properties, including social norms and

This chapter is based on ideas more fully elaborated in Herbert Gintis, *The Bounds of Reason: Game Theory and the Unification of the Behavioral Sciences* (Princeton University Press, 2009).

their psychological correlates, that cannot be derived analytically from lower-level constructs.

Game theory includes four related disciplines: classical, behavioral, epistemic, and evolutionary game theory. Classical game theory is the analytical elaboration of the ideas of John von Neumann and John Nash, widely used in economics. Behavioral game theory is the application of classical game theory to the social psychology of human strategic inter-action, used to show that rational behavior involves deep psychological and sociological principles, including other-regarding behavior and a social epistemology. Epistemic game theory is the application of the modal logic of knowledge and belief to classical game theory, used to integrate the rational actor model with the sociopsychological theory of norms (Gintis 2009a). Evolutionary game theory is a macrolevel analyt-ical apparatus allowing the insights of biological and cultural evolution to be analytically modeled.

The rational actor model, developed in economics and decision theory, is the single most important analytical construct in the behavioral sciences operating at the level of the individual. While gene–culture coevolution-ary theory is a form of "ultimate" explanation that does not predict, the rational actor model provides a "proximate" description of behavior that can be tested in the laboratory and real life, and is the basis of the explana-tory success of economic theory. Game theory makes no sense without the rational actor model, and behavioral disciplines, like sociology and psych-ology, that have abandoned this model have fallen into theoretical dis-array. However, the rational actor model has some obvious shortcomings that must be corrected before the theory can fit synergistically with such fields as social psychology and sociology. One such shortcoming lies in the theory's taking beliefs – or subjective priors, as they are called in rational decision theory – as *sui generis* and not constituted by social processes. In fact, beliefs are the product of social interaction, a fact that undermines the cherished methodological individualism of rational choice theorists. A second shortcoming is that many decisions, including some of the most important and life determining, are not made by a simple maximization subject to constraints, but rather involve considering the experience of others who have made similar choices in the past under similar conditions. Thus, rational decision making inexorably involves *imitation* and *conform-ist bias*. These aspects of rational choice again undermine the methodo-logical individualism of traditional rational decision theorists.

Complexity theory is needed because human society is a complex adap-tive system with *emergent properties* that cannot now, and perhaps never will, be explained starting with more basic units of analysis. The hypothetico-deductive methods of game theory and the rational actor model, and even gene–culture coevolutionary theory, must therefore be complemented by the work of behavioral scientists who deal with society in more macrolevel, interpretive terms and develop insightful schemas that shed light where analytical models cannot penetrate. Anthropological and historical studies fall into this category, as well as macroeconomic policy and comparative economic systems. Agent-based modeling of complex dynamical systems is also useful in dealing with emergent properties of complex adaptive systems.

## 10.1 GENE–CULTURE COEVOLUTION

Because culture is crucial to the evolutionary success of *Homo sapiens*, individual fitness in humans depends on the structure of social life. It follows that human cognitive, affective, and moral capacities are the product of an evolutionary dynamic involving the interaction of genes and culture. This dynamic is known as *gene–culture coevolution* (Cavalli-Sforza and Feldman 1982; Boyd and Richerson 1985; Dunbar 1993; Richerson and Boyd 2004). This coevolutionary process has endowed us with preferences that go beyond the self-regarding concerns emphasized in traditional economic and biological theory and embrace a social epistemology facilitating the sharing of intentionality across minds, as well as such non-self-regarding values as a taste for cooperation, fairness, and retribution, the capacity to empathize, and the ability to value honesty, hard work, piety, toleration of diversity, and loyalty to one's reference group.

The genome encodes information that is used both to construct a new organism and to endow it with instructions for transforming sensory inputs into decision outputs. Because learning is costly and error-prone, efficient information transmission will ensure that the genome encodes all aspects of the organism's environment that are constant, or that change only slowly through time and space. By contrast, environmental conditions that vary rapidly can be dealt with by providing the organism with the capacity to *learn*.

There is an intermediate case, however, that is efficiently handled neither by genetic encoding nor learning. When environmental conditions are positively but imperfectly correlated across generations, each generation acquires valuable information through learning that it cannot transmit genetically to the succeeding generation, because such information is not encoded in the germ line. In the context of such environments, there is a fitness benefit to the transmission of *epigenetic* information concerning the current state of the environment. Such epigenetic information is quite common (Jablonka and Lamb 1995), but achieves its highest and most flexible form in *cultural transmission* in humans and to a considerably lesser extent in other primates (Bonner 1984; Richerson and Boyd 1998). Cultural transmission takes the form of vertical (parents to children), horizontal (peer to peer), and oblique (elder to younger), as in Cavalli-Sforza and Feldman (1981); prestige (higher influencing lower status), as in Henrich and Gil-White (2001); popularity-related, as in Newman et al. (2006); and even random population-dynamic transmission, as in Shennan (1997) and Skibo and Bentley (2003).

The parallel between cultural and biological evolution goes back to Huxley (1955), Popper (1979), and James (1880) – see Mesoudi et al. (2006) for details. The idea of treating culture as a form of epigenetic transmission was pioneered by Richard Dawkins, who coined the term "meme" in *The Selfish Gene* (1976) to represent an integral unit of information that could be transmitted phenotypically. There quickly followed several major contributions to a biological approach to culture, all based on the notion that culture, like genes, could evolve through replication (intergenerational transmission), mutation, and selection.

Cultural elements reproduce themselves from brain to brain and across time, mutate, and are subject to selection according to their effects on the fitness of their carriers (Parsons 1964; Cavalli-Sforza and Feldman 1982). Moreover, there are strong interactions between genetic and epigenetic elements in human evolution, ranging from basic physiology (e.g. the transformation of the organs of speech with the evolution of language) to sophisticated social emotions, including empathy, shame, guilt, and revenge seeking (Zajonc 1980, 1984).

Because of their common informational and evolutionary character, there are strong parallels between genetic and cultural modeling (Mesoudi et al. 2006). Like biological transmission, culture is transmitted from parents to offspring, and like cultural transmission, which is transmitted horizontally to unrelated individuals, so in microbes and many plant species, genes are regularly transferred across lineage boundaries (Jablonka and Lamb 1995; Rivera and Lake 2004; Abbott et al. 2003). Moreover, anthropologists reconstruct the history of social groups by analyzing homologous and analogous cultural traits, much as biologists reconstruct the evolution of species by the analysis of shared characters and homologous DNA (Mace and Pagel 1994). Indeed, the same computer programs developed by biological systematists are used by cultural anthropologists

(Holden 2002; Holden and Mace 2003). In addition, archeologists who study cultural evolution have a similar modus operandi as paleobiologists who study genetic evolution (Mesoudi et al. 2006). Both attempt to reconstruct lineages of artifacts and their carriers. Like paleobiology, archeology assumes that when analogy can be ruled out, similarity implies causal connection by inheritance (O'Brien and Lyman 2000). Like biogeography's study of the spatial distribution of organisms (Brown and Lomolino 1998), behavioral ecology studies the interaction of ecological, historical, and geographical factors that determine distribution of cultural forms across space and time (Smith and Winterhalder 1992).

Dawkins added a fundamental mechanism of epigenetic information transmission in *The Extended Phenotype* (1982), noting that organisms can directly transmit environmental artifacts to the next generation, in the form of such constructs as beaver dams, bee hives, and even social structures (e.g. mating and hunting practices). The phenomenon of a species creating an important aspect of its environment and stably transmitting this environment across generations, known as *niche construction*, is a widespread form of epigenetic transmission (Odling-Smee et al. 2003). Moreover, niche construction gives rise to what might be called a *gene–environment coevolutionary process*, since a genetically induced environmental regularity becomes the basis for genetic selection, and genetic mutations that give rise to mutant niches will survive if they are fitness enhancing for their constructors.

An excellent example of gene–environment coevolution is the honey bee, in which the origin of its eusociality likely lay in the high degree of relatedness fostered by haplodiploidy, but which persists in modern species despite the fact that relatedness in the hive is generally quite low, due to multiple queen matings, multiple queens, queen deaths, and the like (Gadagkar 1991; Seeley 1997; Wilson and Holldobler 2005). The social structure of the hive is transmitted epigenetically across generations, and the honey bee genome is an adaptation to the social structure laid down in the distant past.

Gene–culture coevolution in humans is a special case of gene–environment coevolution in which the environment is culturally constituted and transmitted (Feldman and Zhivotovsky 1992). The key to the success of our species in the framework of the hunter–gatherer social structure in which we evolved is the capacity of unrelated, or only loosely related, individuals to cooperate in relatively large egalitarian groups in hunting and territorial acquisition and defense (Boehm 2000; Richerson and Boyd 2004). While contemporary biological and economic theory have attempted to show that such cooperation can be effected by self-regarding rational agents (Trivers 1971; Alexander 1987; Fudenberg et al. 1994), the conditions under which this is the case are highly implausible even for small groups (Boyd and Richerson 1988; Gintis 2005). Rather, the social environment of early humans was conducive to the development of prosocial traits, such as empathy, shame, pride, embarrassment, and reciprocity, without which social cooperation would be impossible.

Neuroscientific studies exhibit clearly the genetic basis for moral behavior. Brain regions involved in moral judgments and behavior include the prefrontal cortex, the orbitofrontal cortex, and the superior temporal sulcus (Moll et al. 2005). These brain structures are virtually unique to, or most highly developed, in humans and are doubtless evolutionary adaptations (Schulkin 2000). The evolution of the human prefrontal cortex is closely tied to the emergence of human morality (Allman et al. 2002). Patients with focal damage to one or more of these areas exhibit a variety of antisocial behaviors, including the absence of embarrassment, pride, and regret (Beer et al. 2003; Camille 2004), and sociopathic behavior (Miller et al. 1997). There is a likely genetic predisposition underlying sociopathy, and sociopaths comprise 3–4% of the male population, but they account for between 33 and 80% of the population of chronic criminal offenders in the United States (Mednick et al. 1977).

It is clear from this body of empirical information that culture is directly encoded into the human brain, which of course is the central claim of gene–culture coevolutionary theory.

The evolution of the physiology of speech and facial communication is a dramatic example of gene–culture coevolution. The increased social importance of communication in human society rewarded genetic changes that facilitate speech. Regions in the motor cortex expanded in early humans to facilitate speech production. Concurrently, nerves and muscles to the mouth, larynx, and tongue became more numerous to handle the complexities of speech (Jurmain et al. 1997). Parts of the cerebral cortex, Broca's and Wernicke's areas, which do not exist or are relatively small in other primates, are large in humans and permit grammatical speech and comprehension (Binder et al. 1997; Belin et al. 2000).

Adult modern humans have a larynx low in the throat, a position that allows the throat to serve as a resonating chamber capable of a great number of sounds (Relethford 2007). The first hominids that have skeletal structures supporting this laryngeal placement are the *Homo heidelbergensis*, who lived from 800,000 to 100,000 years ago. In addition, the production of consonants requires a short oral cavity, whereas our nearest primate relatives have much too long an oral cavity for this purpose. The position of the hyoid bone, which is a point of attachment for a tongue muscle, developed in *Homo sapiens* in a manner permitting highly precise and flexible tongue movements.

Another indication that the tongue has evolved in hominids to facilitate speech is the size of the hypoglossal canal, an aperture that permits the hypoglossal nerve to reach the tongue muscles. This aperture is much larger in Neanderthals and humans than in early hominids and nonhuman primates (Dunbar 2005). Human facial nerves and musculature have also evolved to facilitate communication. This musculature is present in all vertebrates, but except in mammals, it serves feeding and respiratory functions alone (Burrows 2008). In mammals, this mimetic musculature attaches to the skin of the face, thus permitting the facial communication of such emotions as fear, surprise, disgust, and anger. In most mammals, however, a few wide sheet-like muscles are involved, rendering fine information differentiation impossible, whereas in primates, this musculature divides into many independent muscles with distinct points of attachment to the epidermis, thus permitting higher bandwidth facial communication. Humans have the most highly developed facial musculature by far of any primate species, with a degree of involvement of lips and eyes that is not present in any other species.

This example is quite a dramatic and concrete illustration of the intimate interaction of genes and culture in the evolution of our species.

## 10.2 RATIONAL DECISION THEORY

General evolutionary principles suggest that individual decision making for members of a species can be modeled as optimizing a preference function. Natural selection leads the content of preferences to reflect biological fitness. The principle of expected utility extends this optimization to stochastic outcomes. The resulting model is called the *rational actor model* or *rational decision theory* in economics.

For every constellation of sensory inputs, each decision taken by an organism generates a probability distribution over outcomes, the expected value of which is the *fitness* associated with that decision. Since fitness is a scalar variable, for each constellation of sensory inputs, each possible action the organism might take has a specific fitness value, and organisms whose decision mechanisms are optimized for this environment will choose the available action that maximizes this value. This argument was presented verbally by Darwin (1872) and is implicit in the standard notion of "survival of the fittest," but formal proof is recent (Grafen 1999, 2000, 2002). The case with frequency-dependent (nonadditive genetic) fitness has yet to be formally demonstrated, but the informal arguments are compelling.

Given the state of its sensory inputs, if an organism with an optimized brain chooses action $A$ over action $B$ when both are available, and chooses action $B$ over action $C$ when both are available, then it will also choose action $A$ over action $C$ when both are available. Thus choice consistency follows from basic evolutionary dynamics.

The so-called *rational actor model* was developed in the twentieth century by John von Neumann, Leonard Savage, and many others. The model is often presented as though it applies only when actors possess extremely strong information-processing capacities. In fact, the model depends only on *choice consistency* (Gintis 2009a). When preferences are consistent, they can be represented by a numerical function, often called a *utility function*, which the individual maximizes subject to his subjective beliefs. When consistency extends over lotteries (actions with probabilistic outcomes), then agents act as though they were maximizing expected utility subject to their subjective priors (beliefs concerning the effect of actions on the probability of diverse outcomes).

Four *caveats* are in order. First, individuals do not *consciously* maximize something called "utility," or anything else. Rather, consistent preferences imply that there is an objective function such that the individual's choices can be predicted by maximizing the expected value of the objective function (much as a physical system can be understood by solving a set of Hamiltonian equations, with no implication that physical systems carry out such a set of mathematical operations). Second, individual choices, even if they are self-regarding (e.g. personal consumption) are not necessarily welfare enhancing. For instance, a rational decision maker may smoke cigarettes knowing full well their harmful effects, and even wishing that he preferred not to smoke cigarettes. It is fully possible for a rational decision maker to wish that he had preferences other than those he actually has, and even to transform his preferences accordingly. Third, preferences must have some stability across time to be theoretically useful, but preferences are ineluctably a function of an individual's *current state*, which includes both his physiological state and his current beliefs. Because beliefs can change dramatically and rapidly in response to sensory experience, preferences may be subject to discontinuous change. Finally, beliefs need not be correct nor need they be updated correctly in the face of new evidence, although Bayesian assumptions concerning updating can be made part of consistency in elegant and compelling ways (Jaynes 2003).

The rational actor model is the cornerstone of contemporary economic theory, and in the past few decades has become the heart of the biological modeling of animal behavior (Real 1991; Alcock 1993; Real and Caraco 1986). Economic and biological theory thus have a natural affinity: the choice consistency on which the rational actor model of economic theory depends is rendered plausible by evolutionary theory, and the optimization techniques pioneered in economics are routinely applied and extended by biologists in modeling the behavior of nonhuman organisms.

In a stochastic environment, natural selection will ensure that the brain makes choices that, at least roughly, maximize expected fitness, and hence satisfy the expected utility principle (Cooper 1987). To see this, suppose an organism must choose from action set $X$, where each $x \in X$ determines a lottery that pays $i$ offspring with probability $p_i(x)$, for $i = 0, 1, \ldots, n$. Then the expected number of offspring from this lottery is $\psi(x) = \sum_{j=1}^{n} j p_j(x)$. Let $L$ be a lottery on $X$ that delivers $x_i \in X$ with probability $q_i$ for $i = 1, \ldots, k$. The probability of $j$ offspring given $L$ is then $\sum_{i=1}^{k} q_i p_j(x_i)$, so the expected number of offspring given $L$ is

$$\sum_{j=1}^{n} j \sum_{i=1}^{k} q_i p_j(x) = \sum_{i=1}^{k} q_i \sum_{j=1}^{k} j p_j(x_i) = \sum_{i=1}^{k} q_i \psi(x_i), X, \tag{10.1}$$

which is the expected value theorem with utility function $\psi(\cdot)$.

Evidence from contemporary neuroscience suggests that expected utility maximization is not simply an "as if" story. In fact, the brain's neural circuitry actually makes choices by internally representing the payoffs of various alternatives as neural firing rates, and choosing a maximal such rate (Glimcher 2003; Dorris and Glimcher 2003; Glimcher et al. 2005). Neuroscientists increasingly find that an aggregate decision-making process in the brain synthesizes all available information into a single, unitary value (Parker and Newsome 1998; Schall and Thompson 1999; Glimcher 2003). Indeed, when animals were tested in a repeated-trial setting with variable reward, dopamine neurons appear to encode the difference between the reward that an animal expects to receive and the reward that an animal actually receives on a particular trial (Schultz et al. 1997; Sutton and Barto 2000), an evaluation mechanism that enhances the environmental sensitivity of the animal's decision-making system.

This error-prediction mechanism has the drawback of only seeking local optima (Sugrue et al. 2005). Montague and Berns (2002) address this problem, showing that the orbitofrontal cortex and striatum contains a mechanism for more global predictions that include risk assessment and discounting of future rewards. Their data suggest a decision-making model that is analogous to the famous Black-Scholes options-pricing equation (Black and Scholes 1973).

Perhaps the most pervasive critique of the rational decision model is that put forward by Herbert Simon (1982), holding that because information processing is costly and humans have finite information-processing capacity, individuals *satisfice* rather than *maximize*, and hence are only *boundedly rational*. There is much substance to Simon's premises, especially that of including information-processing costs and limited information in modeling choice behavior, and that of recognizing that the decision on how much information to collect depends on unanalyzed subjective priors at some level (Winter 1971; Heiner 1983). Indeed, from basic information theory and quantum mechanics it follows that *all rationality is bounded*. However, the popular message taken from Simon's work is that we should reject the rational actor model. For instance, the mathematical psychologist D. H. Krantz (1991) asserts, "The normative assumption that individuals *should* maximize *some* quantity may be wrong ... People do and should act as *problem solvers*, not *maximizers*." This is incorrect. In fact, as long as individuals are involved in routine choice (see Section 10.6), and hence have consistent preferences, they can be modeled as maximizing an objective function subject to constraints.

This point is lost on even such capable researchers as Gigerenzer and Selten (2001), who reject the "optimization subject to constraints" method on the grounds that individuals do not in fact solve optimization problems. However, just as the billiards players do not solve differential equations in choosing their shots, so decision makers do not solve Lagrangian equations, even though in both cases we may use such optimization models to describe their behavior.

## 10.3  GAME THEORY

The analysis of living systems includes one concept that is not analytically represented in the natural sciences: that of a *strategic interaction*, in which the behavior of agents is derived by assuming that each is choosing a *best response* to the actions of other agents. The study of systems in which agents choose best responses, and in which such responses evolve dynamically, is called *evolutionary game theory*.

A *replicator* is a physical system capable of drawing energy and chemical building blocks from its environment to make copies of itself. Chemical crystals, such as salt, have this property, but biological replicators have the additional ability to assume a myriad of physical forms based on the highly variable sequencing of its chemical building blocks. Biology studies the dynamics of such complex replicators, using the evolutionary concepts of replication, variation, mutation, and selection (Lewontin 1974).

Biology plays a role in the behavioral sciences much like that of physics in the natural sciences. Just as physics studies the elementary processes that underlie all natural systems, so biology studies the general characteristics of survivors of the process of natural selection. In particular, genetic replicators, the epigenetic environments to which they give rise, and the effect of these environments on gene frequencies account for the characteristics of species, including the development of individual traits and the nature of intraspecific interaction. This does not mean, of course, that behavioral science in any sense *reduces* to biological laws. Just as one cannot deduce the character of natural systems (e.g. the principles of inorganic and organic chemistry, the structure and history of the universe, robotics, plate tectonics) from the basic laws of physics, similarly one cannot deduce the structure and dynamics of complex life forms from basic biological principles. But, just as physical principles inform model creation in the natural sciences, so must biological principles inform all the behavioral sciences.

Within population biology, evolutionary game theory has become a fundamental tool. Indeed, evolutionary game theory is basically population biology with frequency-dependent fitnesses. Throughout much of the twentieth century, classical population biology did not employ a game-theoretic framework (Fisher 1930; Haldane 1932; Wright 1931). However, Moran (1964) showed that Fisher's fundamental theorem, which states that as long as there is positive genetic variance in a population, fitness increases over time, is false when more than one genetic locus is involved. Eshel and Feldman (1984) identified the problem with the population genetic model in its abstraction from mutation. But how do we attach a fitness value to a mutant? Eshel and Feldman (1984) suggested that payoffs be modeled game theoretically on the phenotypic level, and a mutant gene be associated with a strategy in the resulting game. With this assumption, they showed that under some restrictive conditions, Fisher's fundamental theorem could be restored. Their results were generalized by

Liberman (1988), Hammerstein and Selten (1994), Hammerstein (1996), Eshel and Bergman (1998), and others.

The most natural setting for genetic and cultural dynamics is game theoretic. Replicators (genetic and/or cultural) endow copies of themselves with a repertoire of strategic responses to environmental conditions, including information concerning the conditions under which each is to be deployed in response to the character and density of competing replicators. Genetic replicators have been well understood since the rediscovery of Mendel's laws in the early twentieth century. Cultural transmission also apparently occurs at the neuronal level in the brain, in part through the action of *mirror neurons* (Williams et al. 2001; Rizzolatti et al. 2002; Meltzhoff and Decety 2003). Mutations include replacement of strategies by modified strategies, and the "survival of the fittest" dynamic (formally called a *replicator dynamic*) ensures that replicators with more successful strategies replace those with less successful (Taylor and Jonker 1978).

Cultural dynamics, however, do not reduce to replicator dynamics. For one thing, the process of switching from lower- to higher-payoff cultural norms is subject to error, and with some positive frequency, lower-payoff forms can displace higher-payoff forms (Edgerton 1992). Moreover, cultural evolution can involve conformist bias (Henrich and Boyd 1998, 2001; Guzman et al. 2007), as well as oblique and horizontal transmission (Cavalli-Sforza and Feldman 1981; Gintis 2003).

In rational choice theory, choices give rise to probability distributions over outcomes, the expected values of which are the payoffs to the choice from which they arose. Game theory extends this analysis to cases where there are multiple decision makers. In the language of game theory, *players* (or *agents*) are endowed with a set of available *strategies*, and have certain *information* concerning the rules of the game, the nature of the other players and their available strategies, as well as the structure of payoffs. Finally, for each combination of strategy choices by the players, the game specifies a distribution of *payoffs* to the players. Game theory predicts the behavior of the players by assuming each maximizes its preference function subject to its information, beliefs, and constraints (Kreps 1990).

Game theory is a logical extension of evolutionary theory. To see this, suppose there is only one replicator, deriving its nutrients and energy from nonliving sources. The replicator population will then grow at a geometric rate, until it presses upon its environmental inputs. At that point, mutants that exploit the environment more efficiently will outcompete their less-efficient conspecifics, and with input scarcity, mutants will emerge that "steal" from conspecifics who have amassed valuable resources. With the rapid growth of such predators, mutant prey will devise means of avoiding predation, and predators will counter with their own novel predatory capacities. In this manner, strategic interaction is born from elemental evolutionary forces. It is only a conceptual short step from this point to cooperation and competition between cells in a multi-cellular body, between conspecifics who cooperate in social production, between males and females in a sexual species, between parents and off-spring, and between groups competing for territorial control.

Historically, game theory did not emerge from biological consider-ations, but rather from strategic concerns in World War II (von Neumann and Morgenstern 1944; Poundstone 1992). This led to the widespread caricature of game theory as applicable only to static confrontations of rational self-regarding agents possessed of formidable reasoning and information-processing capacity. Developments within game theory in recent years, however, render this caricature inaccurate.

Game theory has become the basic framework for modeling animal behavior (Maynard Smith 1982; Alcock 1993; Krebs and Davies 1997), and thus has shed its static and hyperrationalistic character, in the form of evolutionary game theory (Gintis 2009b). Evolutionary and behavioral game theory do not require the formidable information-processing capaci-ties of classical game theory, so disciplines that recognize that cognition is scarce and costly can make use of game-theoretic models (Young 1998; Gintis 2009b; Gigerenzer and Selten 2001). Thus, agents may consider only a restricted subset of strategies (Winter 1971; Simon 1972), and they may use rule-of-thumb heuristics rather than maximization techniques (Gigerenzer and Selten 2001). Game theory is thus a generalized schema that permits the precise framing of meaningful empirical assertions, but imposes no particular structure on the predicted behavior.

## 10.4 SOCIOPSYCHOLOGICAL THEORY OF NORMS

Complex social systems generally have a division of labor, with distinct social positions occupied by individuals specially prepared for their roles. For instance, a bee hive has workers, drones, queens; and workers can be nurses, foragers, or scouts. Preparation for roles is by gender and larval nutrition. Modern human society has a division of labor characterized by dozens of specialized *roles*, appropriate behavior within which is given by *social norms*, and individuals are *actors* who are motivated to fulfill these roles through a combination of *material incentives* and *normative commitments*.

The centrality of culture in the social division of labor was clearly expressed by Emile Durkheim ([1902] 1933), who stressed that the great multiplicity of roles (which he called *organic solidarity*) required a commonality of beliefs (which he called *collective consciousness*) that would permit the smooth coordination of actions by distinct individuals. This theme was developed by Talcott Parsons (1937), who used his knowledge of economics to articulate a sophisticated model of the interaction between the situation (role) and its inhabitant (actor). The actor/role approach to social norms was filled out by Erving Goffman (1959), among others.

The social role has both normative and positive aspects. On the positive side, the payoffs – rewards and penalties – associated with a social role must provide the appropriate incentives for actors to carry out the duties associated with the role. This requirement is most easily satisfied when these payoffs are independent from the behavior of agents occupying other roles. However, this is rarely the case. In general, social roles are deeply interdependent, and can be modeled as the strategy sets of players in an epistemic game, the payoffs to which are precisely these rewards and penalties, the choices of actors then forming a *correlated equilibrium*, for which the required commonality of beliefs is provided by a society's common culture (Gintis 2009a). This argument provides an analytical link uniting the actor/role framework in sociological theory with game-theoretic models of cooperation in economic theory.

Appropriate behavior in a social role is given by a *social norm* that specifies the duties, privileges, and normal behavior associated with the role. In the first instance, the complex of social norms has a instrumental character devoid of normative content, serving merely as an informational device that coordinates the behavior of rational agents (Lewis 1969; Gauthier 1986; Binmore 2005; Bicchieri 2006). However, in most cases, high-level performance in a social role requires that the actor have a *personal commitment* to role performance that cannot be captured by the self-regarding "public" payoffs associated with the role (Conte and Castelfranchi 1999; Gintis 2009a). This is because (a) actors may have private payoffs that conflict with the public payoffs, inducing them to behave counter to proper performance of their role (e.g. corruption, favoritism, aversion to specific tasks); (b) the signal used to determine the public payoffs may be inaccurate and unreliable (e.g. the performance of a teacher or physician); and (c) the public payoffs required to gain compliance by self-regarding actors may be higher than those required when there is at least partial reliance upon the personal commitment of role incumbents (e.g. it may be less costly to use personally committed rather than purely materially motivated physicians and teachers). In such cases, self-regarding actors who treat social norms purely instrumentally will behave in a socially inefficient manner.

The normative aspect of social roles flows from these considerations. First, to the extent that social roles are considered legitimate by incumbents, they will place an intrinsic positive ethical value on role performance. We may call this the *normative bias* associated with role occupancy (Gintis 2009a). Second, human ethical predispositions include *character virtues*, such as honesty, trustworthiness, promise-keeping, and obedience, that may increase the value of conforming to the duties associated with role incumbency (Gneezy 2005). Third, humans are also predisposed to care about the esteem of others even when there can be no future reputational repercussions (Masclet et al. 2003), and take pleasure in punishing others who have violated social norms (Fehr and Fischbacher 2004). These normative traits by no means contradict rationality (Section 10.2), because individuals trade off these values against material reward, and against each other, just as described in the economic theory of the rational actor (Andreoni and Miller 2002; Gneezy and Rustichini 2000).

The sociopsychological model of norms can thus resolve the contradictions between the sociological and economic approaches to social cooperation, retaining the analytical clarity of game theory and the rational actor model, while incorporating the normative and cultural considerations stressed in psychosocial models of norm compliance.

## 10.5 SOCIETY AS COMPLEX ADAPTIVE SYSTEM

The behavioral sciences advance not only by developing analytical and quantitative models, but by accumulating historical, descriptive, and ethnographic evidence that pays heed to the detailed complexities of life in the sweeping array of wondrous forms that nature reveals to us. Historical contingency is a primary focus for many students of sociology, anthropology, ecology, biology, politics, and even economics. By contrast, the natural sciences have found little use for narrative alongside analytical modeling.

The reason for this contrast between the natural and the behavioral sciences is that *living systems are generally complex, dynamic adaptive systems* with emergent properties that cannot be fully captured in analytical models that attend only to the local interactions. The hypothetico-deductive methods of game theory, the rational actor model, and even gene–culture coevolutionary theory must therefore be complemented by the work of behavioral scientists who adhere to more historical and interpretive traditions, as well as researchers who use agent-based programming techniques to explore the dynamical behavior of approximations to real-world complex adaptive systems.

A *complex system* consists of a large population of similar entities (in our case, human individuals) who interact through regularized channels (e.g. networks, markets, social institutions) with significant stochastic elements, without a system of centralized organization and control (i.e. if there is a state, it controls only a fraction of all social interactions, and itself is a complex system). A complex system is *adaptive* if it undergoes an evolutionary (genetic, cultural, agent-based, or other) process of reproduction, mutation, and selection (Holland 1975). To characterize a system as complex adaptive does not explain its operation, and does not solve any problems. However, it suggests that certain modeling tools are likely to be effective that have little use in a noncomplex system. In particular, the traditional mathematical methods of physics and chemistry must be supplemented by other modeling tools, such as agent-based simulation and network theory.

The stunning success of modern physics and chemistry lies in their ability to avoid or control emergence. The experimental method in natural science is to create highly simplified laboratory conditions, under which modeling becomes analytically tractable. Physics is no more effective than economics or biology in analyzing complex real-world phenomena *in situ*. The various branches of engineering (electrical, chemical, mechanical) are effective because they recreate in everyday life artificially controlled, noncomplex, nonadaptive, environments in which the discoveries of physics and chemistry can be directly applied. This option is generally not open to most behavioral scientists, who rarely have the opportunity of "engineering" social institutions and cultures.

## 10.6 COGNITIVE PSYCHOLOGY AND RATIONAL CHOICE

Cognitive psychology has contributed strongly to our understanding of human choice behavior in recent years. For a considerable period, psychologists used these contributions, improperly, I believe, to mount a sustained attack on the rational actor model. Because there is no general psychological model of deliberative decision making, cognitive psychology has peripheralized decision making in the introductory textbooks. For instance, a widely used text of graduate-level readings in cognitive psychology (Sternberg and Wagner 1999) devotes the *ninth* of eleven chapters to "Reasoning, Judgment, and Decision Making," offering two papers, the first of which shows that human subjects generally fail simple logical inference tasks, and the second shows that human subjects are irrationally swayed by the way a problem is verbally "framed" by the experimenter. A leading undergraduate cognitive psychology text (Goldstein 2005) placed "Reasoning and Decision Making" the *last* of twelve chapters. This includes one paragraph describing the rational actor model, followed by many pages purporting to explain why it is wrong. Summing up a quarter century of psychological research in 1995, Paul Slovic asserted, accurately I believe, that "it is now generally recognized among psychologists that utility maximization provides only limited insight into the processes by which decisions are made" (Slovic 1995, p. 365). "People are not logical," psychologists are fond of saying, "they are psychological."

This conclusion is badly at odds with what we know about the evolution of cognitive capacity. The fitness of an organism depends on the effectiveness of its decision making in a stochastic environment. Effective choice is a function of the organism's state of knowledge, which consists of the information supplied by the sensory inputs that monitor the organism's internal states and its external environment. In relatively simple organisms, the choice environment is primitive and distributed in a decentralized manner over sensory inputs. But, in three separate groups of animals, the craniates (vertebrates and related creatures), arthropods (including insects, spiders, and crustaceans), and cephalopods (squid, octopuses, and other mollusks), a central nervous system with a brain (a centrally located decision-making and control apparatus) evolved. The phylogenetic tree of vertebrates exhibits increasing complexity through time, and increasing metabolic and morphological costs of maintaining brain activity. The brain thus evolved because larger and more complex brains, despite their costs, enhanced the fitness of their carriers. Brains therefore are ineluctably structured to make on balance fitness-enhancing decisions in the face of the various constellations of sensory inputs their bearers commonly experience. The idea that human choice is on balance illogical and irrational, according to this reasoning, is highly implausible.

The human brain shares most of its functions with that of other vertebrates, including the coordination of movement, maintenance of homeostatic bodily functions, memory, attention, processing of sensory inputs, and elementary learning mechanisms. The distinguishing characteristic of the human brain, however, lies in its power in deliberative decision making. Human beings, with the largest brains and the highest proportion of caloric resources devoted to supporting brain activity, are the most competent and effective decision makers in the biosphere. However, this brain power is probably not needed for routine decision making. This may be why the treatment of choice in economics and biology, both of which deal with routine choice, is comfortable with the rational actor model, while the psychology of human choice – which focuses on the *differentia specifica* of human choice, which involve deliberative choice – is not.

Nevertheless, the widespread claims by experimental psychologists that humans are illogical and irrational when dealing with routine choice is simply incorrect, because there are almost always more plausible explanations of the observed behavior than failure of logic or reason. In some cases, the alternative explanation suggests that humans are highly effective decision makers. More often, however, the alternative is compatible with the axioms of rational choice over an appropriate (often nonobvious) choice space, but involve imperfect decision making.

Consider, for instance, Linda the bank teller, a well-known experiment performed by Tversky and Kahneman (1983) that falls into the former category. A young woman Linda is described as politically active in college and highly intelligent, and the subject is asked which of the following two statements is more likely: "Linda is a bank teller" or "Linda is a bank teller and is active in the feminist movement." Many subjects rate the second statement more likely, despite the fact that elementary probability theory asserts that if $p$ implies $q$, then $p$ cannot be more likely than $q$. Since the second statement implies the first, it cannot be more likely than the first.

However, there are several plausible alternatives to faulty reasoning in explaining this behavior. One plausible alternative is based on the notion that in normal conversation, a listener assumes that any information provided by the speaker is relevant to the speaker's message (Grice 1975). Applied to this case, the norms of conversation lead the subject to believe that the experimenter wants Linda's politically active past to be taken adequately into account (Hilton 1995; Wetherick 1995). Moreover, the meaning of such terms as "more likely" or "higher probability" are vigorously disputed even in the professional literature, and may well have a different meaning for the average subject and for the expert. For instance, if the subject were given two piles of identity folders and ask to search through them to find the one belonging to Linda, and one of the piles was "all bank tellers" while the other was "all bank tellers who are active in the feminist movement," the subject might reasonably look through the second (doubtless much smaller) pile first, even though being well aware that there is a "higher probability" that the folder is in the first (much larger) pile rather than the second.

We are indebted to Daniel Kahneman, Amos Tversky, and their colleagues for a series of brilliant papers documenting the various errors intelligent subjects commit in dealing with probabilistic decision making. Subjects systematically underweigh base-rate information in favor of salient and personal examples; they reverse lottery choices when the same lottery is described emphasizing probabilities rather than monetary payoffs, when described in term of losses from a high baseline as opposed to gains from a low baseline; they treat proactive decisions differently from passive decisions even when the outcomes are exactly the same, and when outcomes are described in terms of probabilities as opposed to frequencies (Kahneman et al. 1982; Kahneman and Tversky 2000).

These findings are important for understanding human decision making but they are not a threat to the rational choice model. They are simply performance errors in the form of incorrect beliefs as to how payoffs can be maximized. Decision theory did not exist until this century. Before the contributions of Bernoulli, Savage, von Neumann, and other experts, no creature on Earth knew how to value a lottery. It takes years of study to feel at home with the laws of probability. Moreover, it is costly, in terms of time and effort, to apply these laws even if we know them. Of course, if the stakes are high enough, it is worthwhile to go to the effort, or engage an expert who will do it for you. But generally, as Kahneman and Tversky suggest, we apply a set of heuristics that more or less get the job done (Gigerenzer and Selten 2001). Among the most prominent heuristics is simply *imitation*: decide what class of phenomenon is involved, find out what people "normally do" in that situation, and do it. If there is some mechanism leading to the survival and growth of relatively successful behaviors and if the problem in question recurs with sufficient regularity, the choice-theoretic solution will describe the winner of a dynamic social process of trial, error, and replication through imitation.

While there is no generally accepted model of deliberative decision making in cognitive psychology, research in recent years tends to support a collection of Bayesian models that apply to areas of complex inference. These models are in line with the five principles outlined above.

The idea that rational decision involves choice among a personal library of *small-scale mental models* traces back to Craik (1943). Mental models, according to Craik, have a neural topology that corresponds to a proposed structure of the phenomena they are candidates to represent (Conte and Castelfranchi 1995). Craik's small-scale mental models are accordingly akin to an architect's drawings, to an electronic engineer's schematics, to a molecular biologist's stick-and-ball representation of real molecules, and to a computer scientist's block diagram of a multiprocessor. Cognitive scientists in the Bayesian tradition argue that infants come equipped with a rudimentary repertoire of models for social relationships, for the nature of mind, for language, for causality in the physical world, and for other large world spheres of life with which they must come to terms in the course of maturation. The mind then modifies and chooses among mental models through experience.

This research on neuronal Bayesianism is of central importance to decision theory, as it reasserts the centrality of the rational actor model in an era that is dominated by popular critiques of this model and its rejection by large numbers of behavioral scientists, who tend to offer nothing in its place except perhaps ad hoceries that work for particular cases but are incapable of generalization (Gintis 2009a, ch. 12).

Bayesian models of cognitive inference are increasingly prominent in several areas of cognitive psychology, including animal and human learning (Courville et al. 2006; Tenenbaum et al. 2006; Steyvers et al. 2003; Griffiths and Tenenbaum 2008), visual perception and motor control (Yuille and Kersten 2006; Kording and Wolpert 2006), semantic memory and language processing (Steyvers et al. 2006; Chater and Manning 2006; Xu and Tenenbaum, in press), and social cognition (Baker et al. 2007). For a recent overview of Bayesian models of cognition, see Griffiths et al. (2008). These models are especially satisfying because they bridge the gap between traditional cognitive models that stress symbolic representations and their equally traditional adversaries that stress statistical testing.

Bayesian models are symbolic in that they are predicated upon a repertoire of preexisting models that can be tested, as well as statistical techniques that carry out the testing and provide the feedback through which the underlying models can be chosen and modified.

Bayesian information-processing models may solve the problem of how humans acquire complex understandings of the world given severely underdetermined data. For instance, the spectrum of light waves received in the eye depends both on the color spectrum of the object being observed and the way the object is illuminated. Therefore inferring the object's color is severely underdetermined, yet we manage to consider most objects to have constant color even as the background illumination changes. Brainard and Freeman (1997) show that a Bayesian model solves this problem fairly well, given reasonable subjective priors as to the object's color and the effects of the illuminating spectra on the object's surface.

Several students of developmental learning have stressed that children's learning is similar to scientific hypothesis testing (Carey 1985; Gopnik and Meltzoff 1997), but without offering specific suggestions as to the calculation mechanisms involved. Recent studies suggest that these mechanisms include causal Bayesian networks (Glymour 2001; Gopnik and Schultz 2007; Gopnik and Tenenbaum 2007). One schema, known as constraint-based learning, uses observed patterns of independence and dependence among a set of observational variables experienced under different conditions to work backward in determining the set of causal structures compatible with the set of observations (Pearl 2000; Spirtes et al. 2001). Eight-month-old babies can calculate elementary conditional independence relations well enough to make accurate predictions (Sobel and Kirkham 2007). Two-year-olds can combine conditional independence and hands-on information to isolate causes of an effect, and four-year-olds can design purposive interventions to gain relevant information (Glymour et al. 2001; Schultz and Gopnik 2004). "By age four," observe Gopnik and Tenenbaum (2007), "children appear able to combine prior knowledge about hypotheses and new evidence in a Bayesian fashion" (p. 284). Moreover, neuroscientists have begun studying how Bayesian updating is implemented in neural circuitry (Knill and Pouget 2004).

## 10.7 APPLICATION: SOCIAL NORMS AND CORRELATED EQUILIBRIA

Economists commonly explain social norms as Nash equilibria of an underlying stage game. However, a social norm is better explained as a correlating device with causal effectivity for a correlated equilibrium of the underlying stage game. Whereas the epistemological requirements for rational agents playing Nash equilibria are very stringent and usually implausible, the requirements for a correlated equilibrium amount to the existence of *common priors*, which we interpret as induced by the cultural system of the society in question. In this view, human beings may be modeled as rational agents with special neural circuitry dedicated to reacting to, evaluating, and sustaining social norms. This model of social norms combines the strengths of sociological theory in modeling social norms and the moral behavior of individuals who affirm and comply with such norms, with the analytical power of the economic theory of rational decision making.

When the correlating device, which we will call *the choreographer*, has at least as much information as the players, we need in addition only to posit that individuals obey the social norm when it is costless to do so. When players have some information that is not available to the choreographer (i.e. not all social roles can be fully incentivized), obedience to the social norm requires that individuals have a predisposition to follow the norm even when it is costly to do so. The latter case explains why social norms are associated with other-regarding preferences and provides a basis for a general analysis of corruption in business and government.

Social norms are thus not explained in terms of game theory and Bayesian rationality, but rather are an *emergent property* of human society, which is a complex adaptive system guided by natural selection. Social norms provide a dimension of causal efficacy to social theory, whereas game theory alone recognizes no causal efficacy above the level of individual choice behavior.

Because of the independent causal effectivity of social norms, the standard methodological individualism of classical game theory is untenable. In particular, social norms are predicated upon certain mental predispositions, a *social epistemology*, that is also a product of natural selection. This social epistemology fosters the interpersonal sharing of mental concepts, and justifies the assumption of common priors upon which the identification of Bayesian rationality with correlated equilibrium rests.

For instance, automobile traffic is not regulated as a Nash equilibrium, but rather a correlated equilibrium in which the choreographer is a system of lights and stop signs, augmented by the traffic conventions that condition behavior on the temporal and spatial positions of automobiles. Similarly, in the hawk–dove game, where the unique Nash equilibrium is usually extremely inefficient, by adding the so-called "bourgeois" social norm (which says that the player who first occupies a desired location behaves like a hawk and the player who confronts such an incumbent acts like a dove) leads to social efficiency, we achieve almost perfect efficiency. In these cases, we may assume players are self-regarding but share a common prior as to the nature of the correlating device. A third example is the social norm "a police officer must not accept a bribe from a driver in deciding whether to issue a traffic ticket." Assuming the driver's behavior is known only to the driver and the police officer in question, the social norm is effective only if the officer is willing to sacrifice personal gain in favor of complying with the norm. The correlating device that maintains the reputation of the police force in this case is the social norm of refusing bribes, compliance with which may be high when each officer considers the norm legitimate, perhaps because the norm is socially beneficial and other officers appear to comply with the norm.

The social norm as choreographer has three attractive properties lacking in the social norm as Nash equilibrium. First, the conditions under which rational agents play Nash equilibria are generally complex and implausible (e.g. that all players share a common conjecture concerning

Bob

Alice    *l*        *r*

|   | *l* | *r* |
|---|-----|-----|
| *u* | 5,1 | 0,0 |
| *d* | 4,4 | 1,5 |

Figure 10.1. Alice and Bob try to coordinate.

the strategy choice of each player), whereas rational agents with a com-mon prior canonically play a correlated equilibrium.

Second, the social norms as Nash equilibria approach cannot explain why compliance with social norms is often based on other-regarding moral preferences in which agents choose to sacrifice some personal gain to comply with a social norm. We can explain this association between norms and morality in terms of the incomplete information possessed by the choreographer. Morality, in this view, is doing the right thing even if no one is looking – as long as the cost of so doing is not excessive. Because the motivation to behave morally depends on cost and the gen-eral level of compliance, moral behavior may characterize some but not all social equilibria.

Finally, there are many more correlated equilibria than Nash equilibria in most games. Some of these equilibria are Pareto-superior to all Nash equilibria and/or allow a distribution of payoffs among players that is unavailable with Nash equilibria.

Consider, for instance, a game played by Alice and Bob, with normal form matrix shown in Figure 10.1. There are two Pareto-efficient pure strategy equilibria: (1,5) and (5,1). There is also a mixed strategy equilib-rium with payoffs (2.5,2.5), in which Alice plays *u* with probability 0.5 and Bob plays *l* with probability 0.5.

However, if the players can jointly observe a correlating device that sig-nals *ul* and *dr*, each with probability 1/2, Alice and Bob can then achieve the payoff (3,3) by obeying the correlating device, i.e. by playing (*u*, *l*) if they see *ul* and playing (*d*, *r*) if they see *dr*. Note that this is a Nash equi-librium of a larger game in which the correlating device moves first.

A more general correlated equilibrium for this game can be constructed as developed below. This equilibrium is sufficiently sophisticated that I will drop the charade that there just happens to exist a highly complex correlating device to which both Alice and Bob just happen to be attuned and jointly motivated to follow. Rather, I will stress that such a device must be causally instantiated in society, and the players must have the appropriate psychological machinery to recognize that the characteristics of this device are common knowledge between them. To dramatize this, I have called the correlating device a *choreographer*.

Consider a choreographer who would like to direct Alice to play $d$ and Bob to play $l$ so the joint payoff (4,4) could be realized. The problem is that if Alice obeys the choreographer, then Bob has an incentive to choose $r$, giving him a payoff of 5 instead of 4. Similarly, if Bob obeys the choreographer, then Alice has an incentive to choose $u$, giving her a payoff of 5 instead of 4. The choreographer must therefore be more sophisticated.

Suppose the choreographer has three states. In $\omega_1$, which occurs with probability $\alpha_1$, he advises Alice to play $u$ and Bob to play $l$. In $\omega_2$, which occurs with probability $\alpha_2$, the choreographer advises Alice to play $d$ and Bob to play $l$. In $\omega_3$, which occurs with probability $\alpha_3$, the choreographer advises Alice to play $d$ and Bob to play $r$. We assume Alice and Bob know $\alpha_1$, $\alpha_2$, and $\alpha_3 = 1 - \alpha_1 - \alpha_2$, and it is common knowledge that both have a *normative predisposition* to obey the choreographer unless they can do better by deviating. However, neither Alice nor Bob can observe the state $\omega$ of the choreographer, and each hears only what the choreographer tells them, not what the choreographer tells the other player. We will find the values of $\alpha_1$, $\alpha_2$, and $\alpha_3$ for which the resulting game has a Pareto-efficient correlated equilibrium.

Note that Alice has *knowledge partition* $[\{\omega_1\}, \{\omega_2, \omega_3\}]$, meaning that she knows when $\omega_1$ occurs but cannot tell whether the state is $\omega_2$ or $\omega_3$. This is because she is told to move $u$ only in state $\omega_1$ but to move $d$ in both states $\omega_2$ and $\omega_3$. The conditional probability of $\omega_2$ for Alice given $\{\omega_2, \omega_3\}$ is $p_A(\omega_2) = \alpha_2/(\alpha_2 + \alpha_3)$, and similarly $p_A(\omega_3) = \alpha_3/(\alpha_2 + \alpha_3)$. Note also that Bob has knowledge partition $[\{\omega_3\}, \{\omega_1, \omega_2\}]$ because he is told to move $r$ only at $\omega_3$ but to move $l$ at both $\omega_1$ and $\omega_2$. The conditional probability of $\omega_1$ for Bob given $\{\omega_1, \omega_2\}$ is $p_B(\omega_1) = \alpha_1/(\alpha_1 + \alpha_2)$, and similarly $p_B(\omega_2) = \alpha_2/(\alpha_1 + \alpha_2)$.

When $\omega_1$ occurs, Alice knows that Bob plays $l$, to which Alice's best response is $u$. When $\omega_2$ or $\omega_3$ occurs, Alice knows that Bob is told $l$ by the choreographer with probability $p_A(\omega_2)$ and is told $r$ with probability $p_A(\omega_3)$. Thus, despite the fact that Bob plays only pure strategies, Alice knows she effectively faces the mixed strategy $l$ played with probability $\alpha_2/(\alpha_2 + \alpha_3)$ and $r$ played with probability $\alpha_3/(\alpha_2 + \alpha_3)$. The payoff to $u$ in this case is $5\alpha_2/(\alpha_2 + \alpha_3)$, and the payoff to $d$ is $4\alpha_2/(\alpha_2 + \alpha_3) + \alpha_3/(\alpha_2 + \alpha_3)$. If $d$ is to be a best response, we must thus have $\alpha_1 + 2\alpha_2 \leq 1$.

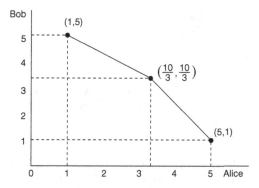

Figure 10.2. Alice, Bob, and the choreographer.

Turning to the conditions for Bob, when $\omega_3$ occurs, Alice plays $d$ so Bob's best response is $r$. When $\omega_1$ or $\omega_2$ occurs, Alice plays $u$ with probability $p_B(\omega_1)$ and $d$ with probability $p_B(\omega_2)$. Bob chooses $l$ when $\alpha_1 + 4\alpha_2 \geq 5\alpha_2$. Thus, any $\alpha_1$ and $\alpha_2$ that satisfy $1 \geq \alpha_1 + 2\alpha_2$ and $\alpha_1 \geq \alpha_2$ permit a correlated equilibrium. Another characterization is $1 - 2\alpha_2 \geq \alpha_1 \geq \alpha_2 \geq 0$.

What are the Pareto-optimal choices of $\alpha_1$ and $\alpha_2$? Because the correlated equilibrium involves $\omega_1 \to (u, l)$, $\omega_2 \to (d, l)$, and $\omega_3 \to (d, r)$, the payoffs to $(\alpha_1, \alpha_2)$ are $\alpha_1(5,1) + \alpha_2(4,4) + (1 - \alpha_1 - \alpha_2)(1,5)$, which simplifies to $(1 + 4\alpha_1 + 3\alpha_2, 5 - 4\alpha_1 - \alpha_2)$, where $1 - 2\alpha_2 \geq \alpha_1 \geq \alpha_2 \geq 0$. This is a linear programming problem. It is easy to see that either $\alpha_1 = 1 - 2\alpha_2$ or $\alpha_1 = \alpha_2$ and $0 \leq \alpha_2 \leq 1/3$. The solution is shown in Figure 10.2.

The pair of straight lines connecting $(1,5)$ to $(10/3,10/3)$ to $(5,1)$ is the set of Pareto-optimal points. Note that the symmetric point $(10/3,10/3)$ corresponds to $\alpha_1 = \alpha_2 = \alpha_3 = 1/3$.

## 10.8 CONCLUSION

I have proposed five analytical tools that together serve to provide a common basis for the behavioral sciences. These are gene–culture coevolution, the sociopsychological theory of norms, game theory, the rational actor model, and complexity theory. While there are doubtless formidable scientific issues involved in providing the precise articulations between these tools and the major conceptual tools of the various disciplines, as exhibited, for instance, in harmonizing the sociopsychological theory of norms and repeated game theory (Gintis 2009a), these intellectual issues are likely to be dwarfed by the sociological issues surrounding the semifeudal nature of the modern behavioral disciplines.

## REFERENCES

Abbott, R. J., J. K. James, R. I. Milne, and A. C. M. Gillies (2003) "Plant Introductions, Hybridization and Gene Flow," *Philosophical Transactions of the Royal Society B* 358: 1123–1132.

Alcock, John (1993) *Animal Behavior: An Evolutionary Approach* (Sunderland, MA: Sinauer).

Alexander, R. D. (1987) *The Biology of Moral Systems* (New York: Aldine).

Allman, J., A. Hakeem, and K. Watson (2002) "Two Phylogenetic Specializations in the Human Brain," *Neuroscientist* 8: 335–346.

Andreoni, James and John H. Miller (2002) "Giving according to GARP: An Experimental Test of the Consistency of Preferences for Altruism," *Econometrica* 70,2: 737–753.

Baker, C. L., Joshua B. Tenenbaum, and R. R. Saxe (2007) "Goal Inference as Inverse Planning," 2007. Proceedings of the 29th Annual Meeting of the Cognitive Science Society.

Beer, J. S., E. A. Heerey, D. Keltner, D. Skabini, and R. T. Knight (2003) "The Regulatory Function of Self-Conscious Emotion: Insights from Patients with Orbitofrontal Damage," *Journal of Personality and Social Psychology* 65: 594–604.

Belin, P., R. J. Zatorre, P. Lafaille, P. Ahad, and B. Pike (2000) "Voice-Selective Areas in Human Auditory Cortex," *Nature* 403: 309–312.

Bicchieri, Cristina (2006) *The Grammar of Society: The Nature and Dynamics of Social Norms* (Cambridge University Press).

Binder, J. R., J. A. Frost, T. A. Hammeke, R. W. Cox, S. M. Rao, and T. Prieto (1997) "Human Brain Language Areas Identified by Functional Magnetic Resonance Imaging," *Journal of Neuroscience* 17: 353–362.

Binmore, Kenneth G. (2005) *Natural Justice* (Oxford University Press).

Black, Fisher and Myron Scholes (1973) "The Pricing of Options and Corporate Liabilities," *Journal of Political Economy* 81: 637–654.

Boehm, Christopher (2000) *Hierarchy in the Forest: The Evolution of Egalitarian Behavior* (Cambridge, MA: Harvard University Press).

Bonner, John Tyler (1984) *The Evolution of Culture in Animals* (Princeton University Press).

Boyd, Robert and Peter J. Richerson (1985) *Culture and the Evolutionary Process* (University of Chicago Press).

(1988) "The Evolution of Reciprocity in Sizable Groups," *Journal of Theoretical Biology* 132: 337–356.

Brainard, D. H. and W. T. Freeman (1997) "Bayesian Color Constancy," *Journal of the Optical Society of America A* 14: 1393–1411.

Brown, J. H. and M. V. Lomolino (1998) *Biogeography* (Sunderland, MA: Sinauer).

Burrows, Anne M. (2008) "The Facial Expression Musculature in Primates and Its Evolutionary Significance," *BioEssays* 30,3: 212–225.

Camille, N. (2004) "The Involvement of the Orbitofrontal Cortex in the Experience of Regret," *Science* 304: 1167–1170.

Carey, Susan (1985) *Conceptual Change in Childhood* (Cambridge, MA: MIT Press).

Cavalli-Sforza, Luigi Luca and Marcus W. Feldman (1981) *Cultural Transmission and Evolution* (Princeton University Press).

(1982) "Theory and Observation in Cultural Transmission," *Science* 218: 19–27.

Chater, N. and C. Manning (2006) "Probabilistic Models of Language Processing and Acquisition," *Trends in Cognitive Sciences* 10: 335–344.

Conte, Rosaria and Cristiano Castelfranchi (1995) *Cognitive and Social Action* (University College of London Press).

(1999) "From Conventions to Prescriptions. Towards an Integrated View of Norms," *Artificial Intelligence and Law* 7: 323–340.

Cooper, W. S. (1987) "Decision Theory as a Branch of Evolutionary Theory," *Psychological Review* 4: 395–411.

Courville, Aaron C., Nathaniel D. Daw, and David S. Touretzky (2006) "Bayesian Theories of Conditioning in a Changing World," *Trends in Cognitive Sciences* 10: 294–300.

Craik, Kenneth (1943) *The Nature of Explanation* (Cambridge University Press).

Darwin, Charles (1872) *The Origin of Species by Means of Natural Selection*, 6th edition (London: John Murray).

Dawkins, Richard (1976) *The Selfish Gene* (Oxford University Press).

(1982) *The Extended Phenotype: The Gene as the Unit of Selection* (New York: W. H. Freeman).

Dorris, Michael C. and Paul W. Glimcher (2003) "Monkeys as an Animal Model of Human Decision Making during Strategic Interactions." Under submission.

Dunbar, R. M. (1993) "Coevolution of Neocortical Size, Group Size and Language in Humans," *Behavioral and Brain Sciences* 16,4: 681–735.

(2005) *The Human Story* (New York: Faber & Faber).

Durkheim, Emile ([1902] 1933) *The Division of Labor in Society* (New York: Free Press).

Edgerton, Robert B. (1992) *Sick Societies: Challenging the Myth of Primitive Harmony* (New York: Free Press).

Eshel, Ilan and Aviv Bergman (1998) "Long-Term Evolution, Short-Term Evolution, and Population Genetic Theory," *Journal of Theoretical Biology* 191: 391–396.

Eshel, Ilan and Marcus W. Feldman (1984) "Initial Increase of New Mutants and Some Continuity Properties of ESS in Two Locus Systems," *American Naturalist* 124: 631–640.

Fehr, Ernst and Urs Fischbacher (2004) "Third Party Punishment and Social Norms," *Evolution & Human Behavior* 25: 63–87.

Feldman, Marcus W. and Lev A. Zhivotovsky (1992) "Gene–culture Coevolution: Toward a General Theory of Vertical Transmission," *Proceedings of the National Academy of Sciences* 89 (December): 11935–11938.

Fisher, Ronald A. (1930) *The Genetical Theory of Natural Selection* (Oxford: Clarendon Press).

Fudenberg, Drew, David K. Levine, and Eric Maskin (1994) "The Folk Theorem with Imperfect Public Information," *Econometrica* 62: 997–1039.

Gadagkar, Raghavendra (1991) "On Testing the Role of Genetic Asymmetries Created by Haplodiploidy in the Evolution of Eusociality in the Hymenoptera," *Journal of Genetics* 70,1 (April): 1–31.

Gauthier, David (1986) *Morals by Agreement* (Oxford: Clarendon Press).

Gigerenzer, Gerd and Reinhard Selten (2001) *Bounded Rationality* (Cambridge, MA: MIT Press).

Gintis, Herbert (2003) "Solving the Puzzle of Human Prosociality," *Rationality and Society* 15,2 (May): 155–187.

    (2005) "Behavioral Game Theory and Contemporary Economic Theory," *Analyze & Kritik* 27,1: 48–72.

    (2009a) *The Bounds of Reason: Game Theory and the Unification of the Behavioral Sciences* (Princeton: Princeton University Press).

    (2009b) *Game Theory Evolving*, 2nd edition (Princeton University Press).

Glimcher, Paul W. (2003) *Decisions, Uncertainty, and the Brain: The Science of Neuroeconomics* (Cambridge, MA: MIT Press).

Glimcher, Paul W., Michael C. Dorris, and Hannah M. Bayer (2005) "Physiological Utility Theory and the Neuroeconomics of Choice." Center for Neural Science, New York University.

Glymour, Alison, D. M. Sobel, L. Schultz, and C. Glymour (2001) "Causal Learning Mechanism in Very Young Children: Two- Three- and Four-Year-Olds Infer Causal Relations from Patterns of Variation and Covariation," *Developmental Psychology* 37,50: 620–629.

Glymour, C. (2001) *The Mind's Arrows: Bayes Nets and Graphical Causal Models in Psychology* (Cambridge: MIT Press).

Gneezy, Uri (2005) "Deception: The Role of Consequences," *American Economic Review* 95,1 (March): 384–394.

Gneezy, Uri and Aldo Rustichini (2000) "A Fine Is a Price," *Journal of Legal Studies* 29: 1–17.

Goffman, Erving (1959) *The Presentation of Self in Everyday Life* (New York: Anchor).

Goldstein, E. Bruce (2005) *Cognitive Psychology: Connecting Mind, Research, and Everyday Experience* (New York: Wadsworth).

Gopnik, Alison and Andrew Meltzoff (1997) *Words, Thoughts, and Theories* (Cambridge, MA: MIT Press).

Gopnik, Alison and L. Schultz (2007) *Causal Learning, Psychology, Philosophy, and Computation* (Oxford University Press).

Gopnik, Alison and Joshua B. Tenenbaum (2007) "Bayesian Networks, Bayesian Learning and Cognitive Development," *Developmental Studies* 10,3: 281–287.

Grafen, Alan (1999) "Formal Darwinism, the Individual-as-Maximizing-Agent: Analogy, and Bet-Hedging," *Proceedings of the Royal Society B* 266: 799–803.

(2000) "Developments of Price's Equation and Natural Selection under Uncertainty," *Proceedings of the Royal Society B* 267: 1223–1227.

(2002) "A First Formal Link between the Price Equation and an Optimization Program," *Journal of Theoretical Biology* 217: 75–91.

Grice, H. P. (1975) "Logic and Conversation," in Donald Davidson and Gilbert Harman (eds.) *The Logic of Grammar* (Encino, CA: Dickenson), pp. 64–75.

Griffiths, Thomas L., Charles Kemp, and Joshua B. Tenenbaum (2008) "Bayesian Models of Cognition." University of California, Berkeley.

Griffiths, Thomas L. and Joshua B. Tenenbaum (2008) "From Mere Coincidences to Meaningful Discoveries," *Cognition* 103: 180–226.

Guzman, R. A., Carlos Rodriguez Sickert, and Robert Rowthorn (2007) "When in Rome Do as the Romans Do: The Coevolution of Altruistic Punishment, Conformist Learning, and Cooperation," *Evolution and Human Behavior* 28: 112–117.

Haldane, J. B. S. (1932) *The Causes of Evolution* (London: Longmans, Green & Co.).

Hammerstein, Peter (1996) "Darwinian Adaptation, Population Genetics and the Streetcar Theory of Evolution," *Journal of Mathematical Biology* 34: 511–532.

Hammerstein, Peter and Reinhard Selten (1994) "Game Theory and Evolutionary Biology," in Robert J. Aumann and Sergiu Hart (eds.) *Handbook of Game Theory with Economic Applications* (Amsterdam: Elsevier), pp. 929–993.

Heiner, Ronald A. (1983) "The Origin of Predictable Behavior," *American Economic Review* 73,4: 560–595.

Henrich, Joseph and Robert Boyd (1998) "The Evolution of Conformist Transmission and the Emergence of Between-Group Differences," *Evolution and Human Behavior* 19: 215–242.

(2001) "Why People Punish Defectors: Weak Conformist Transmission Can Stabilize Costly Enforcement of Norms in Cooperative Dilemmas," *Journal of Theoretical Biology* 208: 79–89.

Henrich, Joseph and Francisco Gil-White (2001) "The Evolution of Prestige: Freely Conferred Status as a Mechanism for Enhancing the Benefits of Cultural Transmission," *Evolution and Human Behavior* 22: 165–196.

Hilton, Denis J. (1995) "The Social Context of Reasoning: Conversational Inference and Rational Judgment," *Psychological Bulletin* 118,2: 248–271.

Holden, C. J. (2002) "Bantu Language Trees Reflect the Spread of Farming across Sub-Saharan Africa: A Maximum-Parsimony Analysis," *Proceedings of the Royal Society B* 269: 793–799.

Holden, C. J. and Ruth Mace (2003) "Spread of Cattle Led to the Loss of Matrilineal Descent in Africa: A Coevolutionary Analysis," *Proceedings of the Royal Society B* 270: 2425–2433.

Holland, John H. (1975) *Adaptation in Natural and Artificial Systems* (Ann Arbor: University of Michigan Press).

Huxley, Julian S. (1955) "Evolution, Cultural and Biological," *Yearbook of Anthropology*: 2–25.

Jablonka, Eva and Marion J. Lamb (1995) *Epigenetic Inheritance and Evolution: The Lamarckian Dimension* (Oxford University Press).

James, William (1880) "Great Men, Great Thoughts, and the Environment," *Atlantic Monthly* 46: 441–459.

Jaynes, E. T. (2003) *Probability Theory: The Logic of Science* (Cambridge University Press).

Jurmain, Robert, Harry Nelson, Lynn Kilgore, and Wenda Travathan (1997) *Introduction to Physical Anthropology* (Cincinatti: Wadsworth Publishing Co.).

Kahneman, Daniel, Paul Slovic, and Amos Tversky (1982) *Judgment under Uncertainty: Heuristics and Biases* (Cambridge University Press).

Kahneman, Daniel and Amos Tversky (2000) *Choices, Values, and Frames* (Cambridge University Press).

Knill, D. and A. Pouget (2004) "The Bayesian Brain: the Role of Uncertainty in Neural Coding and Computation," *Trends in Cognitive Psychology* 27,12: 712–719.

Kording, K. P. and D. M. Wolpert (2006) "Bayesian Decision Theory in Sensorimotor Control," *Trends in Cognitive Sciences* 10: 319–326.

Krantz, D. H. (1991) "From Indices to Mappings: The Representational Approach to Measurement," in D. Brown and J. Smith (eds.) *Frontiers of Mathematical Psychology* (Cambridge University Press), pp. 1–52.

Krebs, John R. and Nicholas B. Davies (1997) *Behavioral Ecology: An Evolutionary Approach*, 4th edn. (Oxford: Blackwell Science).

Kreps, David M. (1990) *A Course in Microeconomic Theory* (Princeton University Press).

Lewis, David (1969) *Conventions: A Philosophical Study* (Cambridge, MA: Harvard University Press).

Lewontin, Richard C. (1974) *The Genetic Basis of Evolutionary Change* (New York: Columbia University Press).

Liberman, Uri (1988) "External Stability and ESS Criteria for Initial Increase of a New Mutant Allele," *Journal of Mathematical Biology* 26: 477–485.

Mace, Ruth and Mark Pagel (1994) "The Comparative Method in Anthropology," *Current Anthropology* 35: 549–564.

Masclet, David, Charles Noussair, Steven Tucker, and Marie-Claire Villeval (2003) "Monetary and Nonmonetary Punishment in the Voluntary Contributions Mechanism," *American Economic Review* 93,1 (March): 366–380.

Maynard Smith, John (1982) *Evolution and the Theory of Games* (Cambridge University Press).

Mednick, S. A., L. Kirkegaard-Sorenson, B. Hutchings, J. Knop, R. Rosenberg, and F. Schulsinger (1977) "An Example of Bio-social Interaction Research: The Interplay of Socio-environmental and Individual Factors in the Etiology of Criminal Behavior," in S. A. Mednick and K. O. Christiansen (eds.) *Biosocial Bases of Criminal Behavior* (New York: Gardner Press), pp. 9–24.

Meltzhoff, Andrew N. and J. Decety (2003) "What Imitation Tells Us about Social Cognition: A Rapprochement between Developmental Psychology and Cognitive Neuroscience," *Philosophical Transactions of the Royal Society B* 358: 491–500.

Mesoudi, Alex, Andrew Whiten, and Kevin N. Laland (2006) "Towards a Unified Science of Cultural Evolution," *Behavioral and Brain Sciences* 29: 329–383.

Miller, B. L., A. Darby, D. F. Benson, J. L. Cummings, and M. H. Miller (1997) "Aggressive, Socially Disruptive and Antisocial Behaviour Associated with Fronto-temporal Dementia," *British Journal of Psychiatry* 170: 150–154.

Moll, Jorge, Roland Zahn, Ricardo di Oliveira-Souza, Frank Krueger, and Jordan Grafman (2005) "The Neural Basis of Human Moral Cognition," *Nature Neuroscience* 6 (October): 799–809.

Montague, P. Read and Gregory S. Berns (2002) "Neural Economics and the Biological Substrates of Valuation," *Neuron* 36: 265–284.

Moran, P. A. P. (1964) "On the Nonexistence of Adaptive Topographies," *Annals of Human Genetics* 27: 338–343.

Newman, Mark, Albert-Laszlo Barabasi, and Duncan J. Watts (2006) *The Structure and Dynamics of Networks* (Princeton University Press).

O'Brien, M. J. and R. L. Lyman (2000) *Applying Evolutionary Archaeology* (New York: Kluwer Academic).

Odling-Smee, F. John, Kevin N. Laland, and Marcus W. Feldman (2003) *Niche Construction: The Neglected Process in Evolution* (Princeton University Press).

Parker, A. J. and William T. Newsome (1998) "Sense and the Single Neuron: Probing the Physiology of Perception," *Annual Review of Neuroscience* 21: 227–277.

Parsons, Talcott (1937) *The Structure of Social Action* (New York: McGraw-Hill).

(1964) "Evolutionary Universals in Society," *American Sociological Review* 29,3 (June): 339–357.

Pearl, J. (2000) *Causality* (New York: Oxford University Press).

Popper, Karl (1979) *Objective Knowledge: An Evolutionary Approach* (Oxford: Clarendon Press).

Poundstone, William (1992) *Prisoner's Dilemma* (New York: Doubleday).

Real, Leslie A. (1991) "Animal Choice Behavior and the Evolution of Cognitive Architecture," *Science* 253 (30 August): 980–986.

Real, Leslie A. and Thomas Caraco (1986) "Risk and Foraging in Stochastic Environments," *Annual Review of Ecology and Systematics* 17: 371–390.

Relethford, John H. (2007) *The Human Species: An Introduction to Biological Anthropology* (New York: McGraw-Hill).

Richerson, Peter J. and Robert Boyd (1998) "The Evolution of Ultrasociality," in I. Eibl-Eibesfeldt and F. K. Salter (eds.) *Indoctrinability, Ideology and Warfare* (New York: Berghahn Books), pp. 71–96.

(2004) *Not by Genes Alone* (University of Chicago Press).

Rivera, M. C. and J. A. Lake (2004) "The Ring of Life Provides Evidence for a Genome Fusion Origin of Eukaryotes," *Nature* 431: 152–155.

Rizzolatti, G., L. Fadiga, L. Fogassi, and V. Gallese (2002) "From Mirror Neurons to Imitation: Facts and Speculations," in Andrew N. Meltzhoff and Wolfgang Prinz (eds.) *The Imitative Mind: Development, Evolution and Brain Bases* (Cambridge University Press), pp. 247–266.

Schall, J. D. and K. G. Thompson (1999) "Neural Selection and Control of Visually Guided Eye Movements," *Annual Review of Neuroscience* 22: 241–259.

Schulkin, J. (2000) *Roots of Social Sensitivity and Neural Function* (Cambridge, MA: MIT Press).

Schultz, L. and Alison Gopnik (2004) "Causal Learning across Domains," *Developmental Psychology* 40: 162–176.

Schultz, W., P. Dayan, and P. Read Montague (1997) "A Neural Substrate of Prediction and Reward," *Science* 275: 1593–1599.

Seeley, Thomas D. (1997) "Honey Bee Colonies are Group-Level Adaptive Units," *American Naturalist* 150: S22–S41.

Shennan, Stephen (1997) *Quantifying Archaeology* (Edinburgh University Press).

Simon, Herbert (1972) "Theories of Bounded Rationality," in C. B. McGuire and Roy Radner (eds.) *Decision and Organization* (New York: American Elsevier), pp. 161–176.

(1982) *Models of Bounded Rationality* (Cambridge: MIT Press).

Skibo, James M. and R. Alexander Bentley (2003) *Complex Systems and Archaeology* (Salt Lake City: University of Utah Press).

Slovic, Paul (1995) "The Construction of Preference," *American Psychologist* 50,5: 364–371.

Smith, Eric Alden and B. Winterhalder (1992) *Evolutionary Ecology and Human Behavior* (New York: Aldine de Gruyter).

Sobel, D. M. and N. Z. Kirkham (2007) "Bayes Nets and Babies: Infants' Developing Statistical Reasoning Abilities and Their Representations of Causal Knowledge," *Developmental Science* 10,3: 298–306.

Spirtes, P., C. Glymour, and R. Scheines (2001) *Causation, Prediction, and Search* (Cambridge, MA: MIT Press).

Sternberg, Robert J. and Richard K. Wagner (1999) *Readings in Cognitive Psychology* (Belmont, CA: Wadsworth).

Steyvers, Mark, Thomas L. Griffiths, and Simon Dennis (2006) "Probabilistic Inference in Human Semantic Memory," *Trends in Cognitive Sciences* 10: 327–334.

Steyvers, Mark, Joshua B. Tenenbaum, E. J. Wagenmakers, and B. Blum (2003) "Inferring Causal Networks from Observations and Interventions," *Cognitive Science* 27: 453–489.

Sugrue, Leo P., Gregory S. Corrado, and William T. Newsome (2005) "Choosing the Greater of Two Goods: Neural Currencies for Valuation and Decision Making," *Nature Reviews Neuroscience* 6: 363–375.

Sutton, R. and A. G. Barto (2000) *Reinforcement Learning* (Cambridge, MA: MIT Press).

Taylor, Peter and Leo Jonker (1978) "Evolutionarily Stable Strategies and Game Dynamics," *Mathematical Biosciences* 40: 145–156.

Tenenbaum, Joshua B., Thomas L. Griffiths, and Charles Kemp (2006) "Bayesian Models of Inductive Learning and Reasoning," *Trends in Cognitive Science* 10: 309–318.

Trivers, Robert L. (1971) "The Evolution of Reciprocal Altruism," *Quarterly Review of Biology* 46: 35–57.

Tversky, Amos and Daniel Kahneman (1983) "Extensional versus Intuitive Reasoning: The Conjunction Fallacy in Probability Judgment," *Psychological Review* 90: 293–315.

von Neumann, John and Oskar Morgenstern (1944) *Theory of Games and Economic Behavior* (Princeton University Press).

Wetherick, N. E. (1995) "Reasoning and Rationality: A Critique of Some Experimental Paradigms," *Theory & Psychology* 5,3: 429–448.

Williams, J. H. G., Andrew Whiten, T. Suddendorf, and D. I Perrett (2001) "Imitation, Mirror Neurons and Autism," *Neuroscience and Biobehavioral Reviews* 25: 287–295.

Wilson, Edward O. and Bert Holldobler (2005) "Eusociality: Origin and Consequences," *Proceedings of the National Academy of Sciences* 102,38: 13367–13371.

Winter, Sidney G. (1971) "Satisficing, Selection and the Innovating Remnant," *Quarterly Journal of Economics* 85: 237–261.

Wright, Sewall (1931) "Evolution in Mendelian Populations," *Genetics* 6: 111–178.

Xu, F. and Joshua B. Tenenbaum (in press) "Word Learning as Bayesian Inference," *Cognitive Sciences* 10: 301–308.

Young, H. Peyton (1998) *Individual Strategy and Social Structure: An Evolutionary Theory of Institutions* (Princeton University Press).

Yuille, A. and D. Kersten (2006) "Vision as Bayesian Inference: Analysis by Synthesis?" *Trends in Cognitive Sciences* 10: 301–308.

Zajonc, Robert B. (1980) "Feeling and Thinking: Preferences Need No Inferences," *American Psychologist* 35,2: 151–175.

(1984) "On the Primacy of Affect," *American Psychologist* 39: 117–123.

# CHAPTER II

# *From fitness to utility*

Kim Sterelny

## II.I MODELLING AGENCY

This essay develops two themes. One is methodological, focusing on the role of model-based science in the human sciences, and, in particular, on model pluralism. I shall discuss three formally similar but causally distinct models of human agency, and shall argue that all three are important in understanding the interaction of individuals and their social context. Two are variants of the rational actor model derived from economics. The simplest of these is the classic rational economic agent model that treats agents as if they act to maximize their material wealth. I contrast it with another version, developed by Herb Gintis and others (Gintis 2006, 2007, 2009), that combines the formal machinery of game theory with the hypothesis that humans are strong reciprocators, and that the psychology of default co-operation plus revenge explains the uniquely co-operative nature of human social life. The third model is formally similar but different in both parentage and underlying causal assumptions, which derive from evolutionary biology. Human behavioural ecology sees humans as fitness rather than utility maximizers. This makes a difference, as we shall see when I take up the second and substantive theme of this essay: I aim to reveal the changing nature of individual agency in the transition from intimate to complex, stratified societies. I shall suggest that we need multiple models as a consequence of diachronic changes in the nature of agency, not just because of the complexity of individuals' relations with their social world. I begin by contrasting the Gintis model of rational agency with its ancestor, and then contrast both to a formally similar approach with roots in a different discipline, evolutionary biology. I then exploit these contrasts in exploring the changing demands on human decision making in the transition from simpler Pleistocene social worlds to the much more complex ones that followed in the Holocene.

Gintis (often in collaboration with Sam Bowles) has argued that humans can be modelled as utility-maximizing agents, thus keeping the formal and conceptual machinery of rational actor models of agency. But Gintis proposes reshaping the standard economists' version of that idea: the version that models humans as self-interested agents making optimal decisions about individual resource acquisition. Rational economic agents will co-operate when that is in their economic interest – you can do business with them – but they are never altruistic (or spiteful). Gintis follows Robert Frank (Frank 1988) in taking gene–culture co-evolution to have reshaped human psychology in ways that require a shift in modelling strategies. Humans have come to live in social worlds in which fitness depends on co-operative partnerships, and hence we have come to live in social worlds in which fitness depends on reputation. We reap the benefits of co-operative partnerships only if we are of good repute, for that makes it likely that we will be chosen by others for profitable partnerships. Fitness depends on agents seeming to be good, to be trustworthy. The best way of seeming to be good is to be good. Our moral emotions evolved in this selective regime, to be both signals of, and motivations for, our social other-oriented dispositions.

Gintis and his collaborators have enriched Frank's model by linking it to the experimental economics literature on 'strong reciprocity'; a set of experimental results that supposedly show that many humans enter interactions disposed to co-operate if they expect co-operation to be mutual, but also willing to punish free riders (Fehr and Gächter 2002; Fehr and Fischbacher 2003; Gächter and Herrmann 2009). Gintis and his colleagues have also developed a broader account of the psychological foundations of these co-operative dispositions. Gene–culture co-evolution has resulted not just in moral emotions but also in psychological predispositions to internalize norms and act on them; and many norms demand prosocial acts, even at individual cost. Finally, they show that this changed view of agency can still be accommodated within the formal framework of rational choice theory and game theory (Bowles and Gintis 2006; Gintis 2007). The crucial idea is that this changed view of agency can be accommodated by a more pluralist conception of agent utilities. Agents care about their own economic welfare. But we care, not only about our own welfare; we sometimes care about the welfare of others. If there are no constraints on what we can include in a model of an agent's values, rational actor models will be empirically empty. But these theorists show that a more pluralistic view of human goals is neither empirically nor computationally intractable, though it is messy.

What might this messier and more realistic conception of human motivation buy us? Realism need not be a virtue in itself. Rational economic agents are models rather than theories of human agency. Models (in contrast to theories) represent their real-world target systems indirectly.[1] A model is not a partial, incomplete or idealized direct description of the world: it is a fully accurate description of an ideal or fictional system. Model specifications are not schematic or partial descriptions of real-world systems: such specifications do not just suppress irrelevant detail; they specify properties known not to be satisfied by any real-world system. So, for example, many population genetics models are models of infinite populations within which organisms mate at random. No biological populations have such properties. Models are useful to the extent that these ideal systems are importantly similar to real-world target systems, and we can use those similarities to give us explanatory and predictive leverage over real-world systems (Godfrey-Smith 2006; Weisberg 2007; Godfrey-Smith 2009). So, for example, the famous prisoner's dilemma is not a realistic picture of real human interactions. We do not make our choices simultaneously, in complete ignorance of others' choices, with perfect knowledge of the outcomes of those choices, and with those outcomes being fully quarantined from all other social interaction. Even so, it is widely accepted that models of iterated prisoner's dilemmas capture important causal mechanisms in human co-operation and its breakdown. Models can also be very informative when we identify their limits: when we can identify when and why the model–target system similarity breaks down. Again, the prisoner's dilemma literature provides an illustration: these models very clearly show that stable multilateral co-operation requires a different explanation than stable bilateral co-operation.

So there is no point in complaining that unlike rational economic agents, real humans care about more than economic resources, or that we are imperfectly rational in assessing odds, given our information. Model specifications are never true of any real-world system. The crucial question is: in what respects do human agents resemble rational economic agents, and what are the consequences of those resemblances for human social worlds? There is much controversy in philosophy of science about when, and why, a model-based approach to science is appropriate (Wimsatt 2007). But there is a partial consensus that (i) models are important when

---

[1] In developing this view on the role of models in an integrated human science, I have been heavily influenced by John Matthewson, Brett Calcott, Michael Weisberg and, especially, Peter Godfrey-Smith, whose picture of the role of models in science I have taken over.

we are confronted with domains that are complex and causally entangled, for in such cases direct representations are computationally intractable. Models represent target systems through similarity relations between the model and the target. Since models can be similar to targets in different respects, we can consistently use distinct, inconsistent models of a single system. For each model can be similar to the target in different ways. (ii) Models are important exploratory tools in the earlier stages of a discipline's development. For theorists need not specify in advance model–target system similarities. One research strategy is to explore the properties of a model (or of a class of related models) and look for surprising similarities between model dynamics and those of the target system.

I shall explore a version of this strategy in this chapter. Human social worlds *are* complex and causally entangled. But they are also dynamic. For consider how deep human history is, and how profoundly the demographic, social and economic worlds of humans have changed over the last 10,000 years. One hundred thousand years ago, *Homo sapiens* individuals lived in small foraging groups, probably confined to Africa, with technology still limited in its diversity and still changing very slowly. Hierarchy and social complexity were probably minimal, with specialization based on division of labour still in the future. Yet these ancient humans were certainly cultural and social beings. Their technology was limited in its variety, but its production and effective use demanded skill, knowledge and co-operation. Over the past 10,000 years, the geographic, demographic, economic and social worlds of most humans have changed in ways those ancient humans could not imagine. Informational demands have changed, as human social worlds have become larger and more stratified, with human action often being routed through social institutions with specific routines and protocols. The time horizon of decision and planning has lengthened: foragers act in time horizons of hours, days, perhaps a week. Subsistence farmers invest time, labour and energy in activity that will pay off, if at all, months later. Contemporary humans invest in education and pension plans whose payoffs come decades after investment. There are important changes in the organization of motivation, as well. Norms, formal rules, and actions based on institutional role replace or supplement motivation based on direct bonds of affiliation and aversion. Paul Seabright, in particular, has emphasized that the Holocene world increasingly involved trusting interaction with strangers, a major break from the past. The extent of these changes (I shall suggest) makes it likely that we need more than a single model of agency. The nature of agency has changed, as human social worlds were transformed.

In particular, I shall suggest that a dynamic view of agency makes sense of two co-existing approaches to agency. One set of ideas derives from economics and allied disciplines, disciplines which represent agents as effective instrumental reasoners. Human behavioural ecologists, in contrast, have adapted tools from behavioural ecology to model humans as fitness optimizers. Human behavioural ecology seeks to show that apparently puzzling patterns of human social worlds – for example, apparently wasteful food taboos – are in fact consequences of adaptive decision making, once we understand the options agents have, and the costs and benefits they must balance. The formal machineries of utility-maximizing and fitness-optimizing models are very similar. But the quantities optimized are not, and nor, as we shall see, are the mechanisms that explain how actions are shaped to ends.

## 11.2 ECONOMIC AND OTHER AGENTS

In this section and the next I explain the two modelling approaches and their strengths. I then suggest a dynamic that explains the transition from agents best modelled through the lens of behavioural ecology to those best modelled with the tools of economics. It is the economist's picture of agency with which I begin. As noted, rational economic agents care only about their own individual welfare. While this simplifies the variety and idiosyncrasy of human motivation, this is an advantage of these models rather than being a weakness. These simplifying assumptions about motivation give them real predictive and explanatory power. For the simplifying assumption is not arbitrary: real humans might care about more than their own material welfare, but they do care about their welfare, typically according it a high priority. Moreover, the assumption that agents have self-regarding economic preferences enables models to be general, strategic, empirically tractable, and explanatorily salient.

The framework is general, because problems of resource acquisition, and the problem of managing trade-offs between risks and benefits, arise in all human social worlds. Rational economic agent models have their most direct application to markets. But they have been applied to a raft of other resource allocation contexts as well, for example, to explanations of the effect of urbanization on family size. Rational economic agent models are consistent with the interactive, strategic character of individual decision making. In many contexts, the consequences of my choices depend in part on others' choices: if I expect you to block an escaping stag by going left, I should stay put; but if I think you will stay put, I should break left. While we do not always choose well (by our own lights) in strategic contexts, we often do, and that shows that we are not opaque to one another: we must often be able to estimate others' goals and their best path to those goals. Rational economic agents who know that they are interacting with other rational economic agents will typically be translucent to one another. The agents in a local interaction will often be able to estimate the local resource envelope that is available, the relative worth of different goods (which will in part be a function of their scarcity and/or the expense of their production), and the most effective resource acquisition strategies. Their own economic activities will generate as a side effect useful information about what other agents want, and how they might get it. So rational economic agents will often make a decent fist of strategic interactions, just as we do.

Rational economic agent models are empirically tractable. There will always be some doubt about whether a given action has maximized the possible gains. For (first) actions are optimal only given a specified set of alternatives, and (second) the fact that an action happened to turn out well (or badly) might show luck rather than best choice. But unlike subjective utility, the economic returns to action are public. So actual outcomes are measurable in the same units the agent uses to value outcomes. Moreover, since models are constructed to represent decision and outcome in reoccurring situations rather than idiosyncratic ones, in practice it is possible to discriminate between good luck and good choice.

Finally, the appeal to economic utility is explanatorily salient. The idea here is that we rob rational actor models of explanatory power if we incorporate norms and customs, and especially culturally contingent norms and customs, into the agent's conception of value. The importance of enculturation is undeniable, but culturally specific norms should not be incorporated into agent preference structure to explain action. We cannot explain (say) the gerontocentric distribution of sexual access in many traditional aboriginal communities (see Keen 2006) by supposing younger male and female agents to have a normatively shaped preference for such partners. For the origin, spread, persistence and influence of such norms is a crucial *explanatory target*, not an *explanatory resource*. Similarly, the East Asian preference for boy children, even in an environment of female shortage, as revealed in patterns of infanticide, is an explanatory target. In these puzzling cases, we would learn nothing from a model of individual choice which simply factored into agent utility functions a strong preference for boys over girls, even when there is independent evidence that agents do have such preferences. Thus a causal model that depends essentially on the motivational power of norms is explanatory when and only when it includes an independent explanation of the origin and stability of those norms.

Thus the ruthless simplicity of the rational economic agent model is crucial to its power. The key question is whether this framework successfully accounts for human co-operation, given realistic specifications of the size of co-operating groups in human cultures, the levels of co-operation they achieve, the frequencies with which individuals expect to interact, and the costs of directing punishment at free riders. As group size increases, a rational economic agent is less likely to co-operate to secure the benefits of repeated direct reciprocation. For in larger groups, when all else is equal, any two agents are less likely to interact regularly. But perhaps in the right circumstances, even in larger groups, self-interested agents will co-operate because reputation is an important resource. It can pay Alex to help Alfred, even if Alfred is unable to return the favour, and even at some cost. For others may see Alex's action and use it as a cue for positive discrimination in future interactions. A self-interested Antonio might use Alex's helpful act as such a cue, both because Alex has signalled readiness to co-operate and (once this show gets off the ground) because Antonio's own reputation will be improved by his positive treatment of other co-operators.

Gintis, Bowles and many others are profoundly sceptical of this vision of the emergence of co-operation among self-interested, reputation-sensitive agents (there is a particularly illuminating exchange between Gintis and Paul Seabright on one side and Don Ross and Ken Binmore on the other: Binmore 2006; Gintis 2006; Ross 2006a, 2006b; Seabright 2006). Their critique has two elements. First, they argue that there are experimentally constructed target systems – experimental public goods games, ultimatum games, three-player ultimatum games, and the like, which are demonstrably not captured by reputation-sensitive rational economic agent models. For (i) these target systems are structured to exclude the possibility of rationally investing in reputation; and (ii) in an important class of cases, co-operation still emerges. Moreover, these targets are structured in such a way that if agents have non-self-interested, social utilities, there will be a clear and consistent signal of those social utilities. Such signals are indeed seen.

Second, they argue that these models of reputation-mediated co-operation do not capture crucial aspects of co-operation as it emerges in natural communities: there are crucial differences between the fictional systems of interacting, reputation-sensitive but self-interested agents, on the one hand, and these real communities, on the other. In the models, agents are unrealistically well-informed about one another. In the models, reputation tracks actual action too exactly, because in real life, social information is not free. Thus an agent refusing to co-operate with a third party might be defecting, or might be punishing a defector: the difference is important; but it is not obvious to casual inspection, and not available for free. So real agents' information about their social constellation will be both incomplete and somewhat inaccurate. Moreover, these models of prudential co-operation assume that punishment – the exclusion of free riders through negative discrimination – is virtually free. But that is not true of target systems with high levels of co-operation.

### 11.2.1 What does experimental economics show?

In response, defenders of the rational economic agent framework have argued that the experimentally constructed target systems of behavioural economics are themselves models of naturally occurring, socially salient real-world systems,[2] and they are poor models of such systems. In developing this sceptical view of the experimental data supposedly showing that humans are 'strong reciprocators', it is important that defenders of the rational economic agent framework do not take that framework to represent the computational architecture of the mind. Rather, they are models of resilient behavioural dispositions in a central class of decision contexts. These contexts are (a) central to the life prospects of the agent in question and central, through the outcomes of decision, to the social life of the community of which the agent is a member; and (b) stereotypical.

The agent in question (and typically other agents in similar situations) has a history of decision making in similar situations, with similar options and cost–benefit trade-offs. In virtue of the first of these conditions, it is worth developing effective decision heuristics; in virtue of the second, it is possible to do so. From the perspective of cognitive science, we are probably far-from-ideally rational, mixed motivation bundles of kludges. But with appropriate practice, and with appropriate environmental support, in a limited but salient set of contexts, we can marshal heuristics that approximate ideal rationality. In many experimental economics settings, these heuristics do not deliver economically optimal

---

[2] But a model not in the sense of a formal or fictional model, but a model system: in the way rats and pigeons have been model systems for those working on associative learning, or the fruit fly has been for those working on genetics and development.

outcomes. For these experimental contexts have been designed to subvert adaptive decision heuristics, to drive a wedge between what is prudent in reputation-sensitive, repeated-interaction contexts and what is prudent in blind, often one-off contexts. We rely on habits and values which are prudent in many naturally occurring and familiar settings, but these will often not optimize in many experimental settings. Even though they fail to optimize in such experimental settings, perhaps most humans act like rational economic agents in the core business of their daily life, and that of their community. If so, that is vindication enough of rational economic agent modelling.

From this perspective, whether an agent is well-modelled as a rational economic agent does not depend on the subjective, introspectable character of their immediate motivation. It depends on whether in the socially central contexts these motivations – independently of their qualitative or representational character – in fact reliably prompt action that advances the material interests of the agent. In experimental contexts, the 'pro-social' emotions identified by Fehr and Gächter may well lead to a sacrifice of material interest. It by no means follows that they are similarly sacrificing in real-world target contexts. The experimental games provide evidence that the rational economic agent framework does not model the intrinsic computational architecture of humans well. Hence it does not give a context-general model of human decision making. But few defenders of the rational economic agent framework think of it as a model of *thinking*. They see that framework as modelling the outcomes of choices in strategic, repeated, highly salient contexts. And that idea remains live in the face of experimental economics.

### 11.2.2 *Would the prudent but selfish co-operate in multiplayer interactions?*

The core social business of human life includes many co-operative, multi-agent interactions. If these fall outside the scope of rational economic agent models, these models are brutally limited. Ross and Binmore argue that Gintis, Seabright and others overstate the problem of information access and expensive punishment, problems that supposedly undermine prudential co-operation in such interactions. They point out that information flows through a community as a side effect of agents' utilitarian activities. Individuals interact, and observe others interacting, in the course of their ordinary activities. As individuals interact, they leak information about their capacities, habits and dispositions. To find out about Alex and

Alfred, often Antonio does not have to invest in epistemic action; he simply needs to be ready to pick up freely available information. (Such free information is not full information, but it may be full enough.) Moreover, they argue that punishment is typically costless too. In most circumstances, punishment consists of no more than a zero-cost social warning ('I see you') at an incipient social transgression, a signal that is noted and acknowledged as the transgressive act is aborted with no damage done. In support of this interpretation, Don Ross sketches a familiar pub scenario: an agent who seems reluctant to pay for his round is reminded of the standard operating procedure with a just-pointed-enough joke. The incipient freeloader publically takes it as a joke that includes rather than excludes him. But he buys his round, and social amity is restored. Policing is necessary, effective, but because it is pre-emptive, heading defection off rather than punishing after it has occurred, it is cheap.

In my view, Ross and Binmore are largely right about information, but wrong about punishment; once co-operation is established in human groups, information does flow freely, because work is social. We extract, transform and often distribute resources in teams. Close ecological and economic co-operation generates information about one another as a free side effect (Sterelny 2007). Once co-operation is established as a stable feature of human social worlds, agents will often no longer routinely have to invest time and effort in finding out about one another (though in special circumstances, in high-stakes decisions, they may). However, this free access to information changes in the transition to larger and more complex societies in the late Holocene. As social worlds become larger, and as they develop more specialization and social stratification, the automatic flow of information in the village glasshouse dwindles, and increasingly we need to rely on third-party opinion through gossip networks, and on social markers (the right accent, the right clothes). The cost of information goes up, and its reliability falls.

Even so, we do know a lot about other agents. I am more sceptical of their line on punishment. Ross's scenario is indeed familiar, and he is right in thinking that pre-emptive punishment can be low-cost and effective, for no costs on co-operators have yet been imposed by defection. But it is a best-case scenario. It ignores risk costs. Those of us who share Ross's familiarity with norm management at the pub will also recall social groups fracturing, as one agent resents attempted social sanction[3]

---

[3] Often because they do not think they have violated any norm that they recognize – there does seem good evidence that we much resent mistaken punishment.

and tempers flare. At the limit, such irruptions can be lethal (Seabright (2010) details the very high murder rates in co-operative forager societies). Attempted low-cost punishment always risks turning into high-cost, even very high-cost, punishment. Moreover, we need to ask why 'soft' social sanctions are often effective. In part, they are effective because of an implied threat to exclude the agent from a material benefit. Even a rational economic agent may weigh heavily a threat to the flow of beer. But they are also effective because people care about subjective interpersonal rewards, about how others see them. Rational economic agents are relatively impervious to soft sanctions, for they have only an instrumental interest in others' regard.

The strong-reciprocation conception of human agency can be partially reconciled with the rational economic actor model. It may well be that in the intimate world of the Pleistocene foraging band, the cool-hearted, farsighted prudent rational economic maximizer would have co-operated consistently with others in the community. Co-operation was prudent: there were very considerable staghunt and risk-management gains from co-operation. Free-riding would be imprudent: interaction was frequent; information was widely available. Punishment was sometimes cheap (for example, through exclusion in contexts of partner choice). When not cheap, punishment could sometimes be co-opted as an honest signal of quality and commitment. So free riding would be risky. Rational economic maximizers in the Pleistocene might well look just like strong reciprocators. Perhaps that is how our ancestors managed maximization. For as Sarah Hrdy (2009) very vividly illustrates, we were not, and did not evolve from, cool-headed calculators but from hot-headed great apes. So the psychology of strong reciprocation may well have been the only available way of engineering an approximation to rational economic agency from a starting point of hominins with rudimentary but powerful social emotions and poor impulse control. But once the flat and intimate social world of the Pleistocene was displaced by stratified societies in the Holocene, the two models of agency diverge. They probably predict very different patterns in action, as opportunities for free-riding emerge.

Strong-reciprocator psychology is still a version of instrumental rationality. Strong reciprocators have utility functions in which objective costs and benefits are modulated by subjective costs and rewards, as the welfare of others counts to an agent in his or her assessment of the utility of an act. For example, punishment is contingent on perceived cost and impact. Thus if competition is added to the ultimatum game, so the first player can make an offer to several candidate second players, those second players are less willing to punish a low-ball offer by refusing (presumably because the punishment might fail through the low-ball offer being accepted by the rival second player). Strong reciprocators do not punish independently of costs and benefits. So this version of the rational actor model has many of the virtues of the rational economic agent framework. Agent utilities are common across agents and stable over time, for these depend on social and moral emotions that are developmentally stable features of our evolutionary inheritance. As with the rational economic agent model, this view of agency fits with our capacity to manage strategic interaction. We are translucent to one another, because our utilities in part depend on our access to objective, public resources, and in part because they depend on subjective features of agent psychology which are displayed by emotional signals. Importantly, to the extent that these rational social agent models are based on prosocial emotions rather than norm acquisition, social preferences are independent of, and explanatorily prior to, the social phenomena – the networks of human co-operation and pro-cooperation norms – that these models explain. Prosocial emotions are an explanatory resource for a theory of human co-operation (granted a theory of their evolution). They are not just one of the phenomena such theories are meant to explain. Finally, rational social agent models have broad scope, as they are meant to help explain not just a wide range of co-operative interchanges, but also those in which deterrence is important, for example bargaining interactions where the weaker side must have a credible option to walk away from the negotiations. In these models, the social and moral emotions make our threats credible, not just our promises.

The idea, then, is that this model of agency is not as simple as the rational economic agent model, but it will predict human action in a broader range of contexts. As noted above, that may be crucial. For while rational economic agency models might well predict that humans co-operate in the intimate glasshouse of Pleistocene forager bands, it is much less clear that they predict co-operation in the larger, noisier and less intimate worlds of the Holocene.[4] The bottom line: rational actor models can be empirically tractable, general and explanatorily powerful without presupposing that agents have self-interested motivations.

---

[4] And yet the social contract did survive in these larger social worlds, even before the establishment of coercive machineries of early states. I take up this puzzle in Sterelny (forthcoming).

## 11.3 HUMAN BEHAVIOURAL ECOLOGY

There is therefore an impressive case to be made for rational agent models of human action. However, co-existing with these economics-derived models, there is an alternative framework that descends directly from evolutionary biology. Animals sometimes behave in very puzzling ways, engaging in expensive displays, long migrations, sometimes imposing limits on their own seasonal reproduction, sometimes reproducing once, and then dying. Some change sex, others mimic the opposite sex. Behavioural ecologists have developed plausible models showing that such apparently dysfunctional behaviours in fact maximize fitness. Thus behavioural ecologists represent organisms as fitness optimizers, though they optimize subject to constraints imposed by their potential phenotypes, their environment, and the choices of other agents (Krebs and Davies 1997).

Human behavioural ecologists have co-opted this program. Human behavioural ecologists build models of distinctive, central patterns of human action: fertility decisions, food gathering, resource sharing. Individuals optimize. But they optimize subject to constraints imposed by the conditions under which they act and the choices of other agents (Winterhalder and Smith 2000; Laland and Brown 2002; Smith and Winterhalder 2003; Laland 2007). Such models of adaptive decision making have a common structure. They specify the optimal behaviour, given a set of assumptions about the environment, the range of options available, and the impact of choice on fitness. Forager women, for example, face difficult trade-offs over birth spacing, as their nomadic life style and the costs of mobility mean that they cannot care for a second child until the first can travel unassisted (see for example Kaplan 1996; Blurton Jones et al. 2006). So a !Kung pattern of birth spacing is the result of choice constrained by social and physical environment. !Kung women respond both to the objective resource envelope and to the choices that others make. If her social group moves, a !Kung mother must move with it (Blurton Jones 1997a, 1997b).

There is considerable predictive overlap between these three approaches to human behaviour. Seeing humans as strong reciprocators and seeing them as rational economic agents make equivalent predictions when individuals maximize their resources through co-operation, and where it is prudent to control defection by punishment. In turn, these models are equivalent to those from human behavioural ecology in those contexts in which material wealth is the most crucial fuel for fitness. But these are substantive restrictions, so the overlap is partial. There is an obvious question, for example, about the extent to which maximizing wealth maximizes fitness in contemporary western societies. There is a similar issue about forager societies. Eric Alden Smith and his colleagues argue that embodied and social capital are more important to forager success than material wealth, in part because they have little opportunity to accumulate material wealth and in part because skill and social support are so important to their lifeways (Smith et al. 2010).

An impressive and illustrative paradigm of human behavioural ecology is the long debate over male large-game hunting in forager societies (recently reviewed in Gurven and Hill 2009). On one side, it has been argued that male hunting cannot be explained as an essentially economic activity: because it is not profitable enough relative to alternatives or because hunters do not keep enough of these profits. According to this view, hunting is a form of signalling. Their critics argue that the signal hypothesis underestimates the profits of hunting and the extent to which sharing those profits is contingent on return benefits (Bliege Bird et al. 2001; Hawkes and Bird 2002; Gurven and Hill 2009). Once we do the accounting correctly, hunting makes economic sense. Despite these differences in interpretation, both sides presume that agents typically maximize their individual fitness (or perhaps that of their family or local group). Large-game hunting is a long-established, widespread, core activity in many human cultures. Such patterns of action reflect and respond to human selective environments. Neither those that defend the signalling model, nor those sceptical of it, think male hunting is satisfactorily explained by the existence of norms that endorse both hunting, and sharing the proceeds of those hunts, even though such norms exist (Boehm 1999). Their shared assumption is that norms inconsistent with interest are unlikely to establish or be stable, if established.

Human behavioural ecologists portray agents as acting adaptively, given their environments and the options open to them. The idea, though, is not just that some forms of human action are adaptive. Rather, these forms of action occur *because* they are adaptive. There is a systematic connection between environmental challenge and agent response: if we understand the challenge, we can predict and explain the response. How, though, does the fact that (say) hunting success enhances fitness by signalling good genes explain the fact that men hunt? They are not consciously aiming to maximize the number of their grandchildren through such signals. Nor is male hunting a wired-in, innate drive. Few males in contemporary cultures impress women by stopping for roadkill. The use of the same formal models of standard behavioural ecology disguises the fact that the explanatory structure of human behavioural ecology is quite different from that of its parent discipline.

In behavioural ecology simpliciter, adaptive fit depends on population history. Salmon spawn and die because they are playing a game against the world, and given its parameters, they do best by one-shot breeding. River journeys are high-risk activities for large, visible fish in small clear streams. There is no point in saving resources for the future if there is unlikely to be a future, and so salmon maximize their reproductive success by staking all their resources on the first mating effort (Quinn 2005). Ancient salmon that played a different strategy – that reserved resources for a second attempt – were less fit than those who invested more in their first go. The result is a population with little variation or individual flexibility. Salmon would not (I suspect) respond to safe streams by staggering their reproductive effort. The adaptive fit between action and environment depends on the history of the population, as it experiences repeated selective filtering. In contrast to the models of agency that descend from economics, adaptation depends on the 'intelligence' of the lineage, not just that of the individual agent.

The salmon breeding strategy is invariant. But even when populations vary, adaptive responses studied by behavioural ecology often depend on population history. Consider, for example, those species in which we find multiple mating strategies. Powerful males guard resources; other males seek sexual access by mimicking females (Jukema and Piersma 2006). These systems often include individual flexibility: as a male matures, he may switch from a cryptic to a resource-defending strategy. But even though the animals act in different ways in different stages of their life, they continue to conform to a single, though conditional, behavioural rule. Each male cuttlefish (for example) conforms to the strategy: guard if you are large; otherwise imitate being female (Norman et al. 1999). As with salmon investment decisions, this strategy is pervasive in the population as a result of selection on previous populations which probably included other mating strategies. Perhaps some males attempted to defend resources independently of their size; others did not switch when the time was right. The adaptive response of individuals is explained by the repeated filtering of variation that existed in previous populations.

Rational agency models depend on individual adaptive response in explaining the fit between an agent's actions and goals. Human behavioural ecology is intermediate between classic behavioural ecology, where fit depends on the evolutionary plasticity of the lineage, and the intelligent, informed, adaptable individual of economics. Population-level processes do play a central role in explaining human fitness-maximizing strategies. But the models of human behavioural ecology typically presuppose individual adaptive phenotypic plasticity. Let me illustrate its role before turning to that of population-level processes.

Consider a classic of the 'hunting is a signal' literature: Bliege Bird et al. (2001) on Meriam turtle hunting. They argue that turtle hunting is a costly signal of male fitness: its rewards are those of status and esteem, rather than calories. The fitness model may be the same as that of Zahavi on the Arabian babbler (Zahavi 1990). But there is no assumption that the strategy of hunting turtles as a signal has outcompeted other strategies over deep time. The Mer live partly by subsistence and partly in the cash economy, and hunting turtles for status might well be a quite new custom; certainly, they use store-bought equipment in the hunt. Turtle hunters may well be maximizing their fitness, but individual adaptability plays a central role in establishing a match between actor, environment and reward. Human response to the environment owes everything to behavioural plasticity. We adapt through learning, and those capacities often enable agents to adapt in spite of new challenges from the physical, biological and social environment.

As with rational agent models, human behavioural ecologists do not model thinking, certainly not conscious, explicit thinking. We do not consciously, deliberately, maximize fitness. So rather than being models of individual computational architecture, they too are models of resilient behavioural dispositions. The core achievement of these models, when they are successful, is to identify the real options available to the agent, and the costs and benefits imposed by the environment on these options. So these models represent the interaction between choice, current environment and outcome. But while they do not model thinking, they have cognitive presuppositions. They implicitly rely on agents having the ability to track their world. The explanation of !Kung fertility relies on humans having cognitive mechanisms that allow them to assess the causal structure of their environment, and to recognize the likely consequences of their actions. And it presupposes that proximate motivation tracks fitness. Humans do not consciously aim at maximizing their fitness any more than salmon do. But the outcomes we prefer increase fitness; those we avoid reduce fitness. So while !Kung women do not plan to maximize the number of their surviving grandchildren, they do plan to have (say) well-fed and well-behaved children, and those goals (if achieved) covary systematically with their fitness. The proximate targets of action are fitness resources. As we shall see, population-level processes play a key role in stabilizing the connection between these proximate goals and fitness resources, and in providing the informational resources for adaptive action.

## II.4 FROM FITNESS TO UTILITY

Human behavioural ecology models have been very insightful in exploring dynamics of traditional cultures. So, for example, there are models of human life-history evolution and the puzzling fact that despite human infants being extraordinarily helpless, expensive and long dependent, we have relatively short interbirth intervals compared with chimps. There are 'polygyny threshold' models that explore the circumstances in which a women should choose to be a second wife in a multifemale household (Marlowe 2000). Yet while these models have been widely and insightfully applied to human life in small-scale traditional societies, these models have been almost invisible in explaining human action in large-scale worlds (perhaps with the exception of mate choice). Why would that be? If we can insightfully model agents as fitness maximizers in small worlds, why is it so much less useful for large worlds? That is not because small-scale worlds are precultural, that 'biology dominates culture in, but only in, such worlds'. Ethnographers have amply documented the very rich cultural life of historically known foragers and traditional agriculturalists. While there is some controversy about the richness of the symbolic and normative life of the very earliest members of our species, the physical record of rich culture is at least 40,000 plus years old (Klein and Edgar 2002). Moreover, natural selection continues to act in large-scale social worlds. Indeed, if Jared Diamond and others are right, the fate of large-scale social worlds has often depended on differential susceptibility to different diseases (this idea is not new; see Zizsser 1935, Diamond 1998, Crosby 2004). All core human activities are profoundly influenced by both our inherited culture and our inherited biology.

So at this point of this essay, the emphasis changes from the first to the second theme: from methodological issues about models and model choice to the substantive project of revealing the changing nature of individual agency in the transition from intimate to complex, stratified societies. The transition to large-scale social life reshapes the sources of cultural information. Individuals remain adaptable and responsive, but population-level mechanisms change, reshaping the extent to which individual decision making tracks fitness. As a consequence of reshaped cultural learning, proximate motivation is less well tuned to fitness interests. In fluent, practiced decision making, we continue to make effective decisions to get want we want. But what we want less reliably tracks our genetic interests. The propensity of individuals in small-scale worlds to make near-optimal fitness-maximizing decisions depends on their capacity to acquire the information they need and to tune their subjective goals to their objective needs. These mechanisms of individual adaptation are stressed in the transition to large-scale social worlds. As human social worlds shrink in number and variety, but grow in size and vertical complexity, the mix of biological and cultural factors changes. As a consequence of that changed mix, human agents shift from being fitness maximizers (perhaps roughly approximated as resource maximizers) to utility maximizers. For the transition to mass society makes tracking instrumental information more complex, and makes the connection between proximal motivation and fitness more fragile. But despite the increased complexity of these new social worlds, our mechanisms of rational appraisal are intact. In a range of key cases, we continue to make instrumentally effective decisions. But the link between conscious goals and fitness resources is fractured. We less reliably want what our genes need.

### 11.4.1 Population-level mechanisms

Population-level mechanisms are important in explaining the link between decision making and fitness in traditional cultures. Vertical (and near-vertical) cultural inheritance is important for the accumulation of cognitive resources essential for instrumental rationality. Boyd and Richerson are fond of contrasting the grim fate of the Burke and Wills expedition across central Australia (almost all died) with the untroubled survival of the locals, who had the benefits of the accumulated lore of their ancestors. We live in complex and somewhat conflicted social worlds; we often live in risky physical environments; we live by extracting high-value but often heavily defended resources from the biological and physical environment. So effective instrumental reasoning is information hungry. In small-scale human social worlds, this information is assembled gradually, generation by generation, filtered, gradually improved, and transmitted to the next generation with high reliability. Peter Richerson, Robert Boyd and their co-workers have shown this process to be central in guiding adaptive behaviour in traditional social worlds. They developed an array of formal models of transmission, mostly to show that reliable transmission may not require high fidelity learning in specific learning episodes. Redundancy and repetition can compensate for imperfect learning. Yet to make adaptive decisions, agents need more than good information. They need good goals too. Since agents do not (typically) subjectively aim at maximizing fitness, human behavioural ecology presupposes that there is a stable correlation between the proximate desires of agents and fitness: what agents want are fuels for fitness. Children inherit values as well as information from their parents. Maladaptive values are filtered out, just as maladaptive beliefs are.

The transitions to mass society make these filtering mechanisms less effective: children acquire less of their information and less of their values and goals from their parents, and from close associates of their parents who are informational and ideological duplicates of their parents. For as social worlds become larger and more connected, the shape of the cultural transmission network has changed from closed-vertical flow to open, much more oblique and horizontal flow. As social worlds become larger, individuals resemble one another less and less in belief and value: for these reflect not just idiosyncratic individual history but social and occupational role, and place in the social hierarchy. So the difference between vertical and other forms of cultural learning makes an increasing difference to what is acquired. So a hypothesis suggests itself: with the transition to mass societies, there is a radical decline in individual-level heritability of those traits whose development is culture dependent. In traditional societies, children ideologically and informationally resemble their parents, because they inherit – to a significant degree – their values and goals, and their instrumental lore. Seriously maladaptive lore and values are less likely to be rebuilt in the next generation, for the same reason seriously maladaptive genes become rarer.

So individual to individual heritability may well have declined, with increasing social complexity. That is not the only important change in population-level processes. There is a persuasive case for the idea that selection on culturally defined human groups has been a mechanism of potential importance in human evolution. Group selection is powerful when (i) the metapopulation is large, with significant and stable variation between groups; (ii) groups differ from one another in their prospects for survival, and for founding new groups; (iii) groups that bud from parent groups resemble those parents; (iv) the groups themselves are internally homogenous (for otherwise individual selection within groups may preempt group selection). Arguably, the growing importance of culture and cultural life resulted in human metapopulations satisfying these conditions (Sober and Wilson 1998; Richerson and Boyd 2001; Richerson et al. 2003; Boyd et al. 2005; Gintis 2007). Group selection would reinforce individual cultural inheritance: groups which have practices that make them more efficient in accumulating, filtering or transmitting high-utility information would be favoured at the expense of groups less able to accumulate or conserve their informational resources.[5]

Arguably, cultural group selection has played an even more important role in filtering maladaptive norms and values: in ensuring that what people value is good for their reproductive prospects. Gintis makes much of the fact that (most) people internalize the norms of their community: they make the community's values their own, and that influences what they want. Thus, many in the west find the idea of eating dogs not just contrary to the prevailing social norms but repellent, disgusting. Boyd and Richerson point out that there is no local guarantee that the norms established in a community are adaptive for that community or the individuals in it. There need only be a small fraction of zealots to establish a local custom, for they are willing to punish those who do not conform (Boyd and Richerson 1992; Boyd et al. 2005). Once established, it then becomes normalized for the succeeding generations. If the zealots are zealous enough, seriously maladaptive norms and customs can become locally entrenched. Boyd and Richerson suggest that such maladaptive norms are weeded out by group selection. Communities with maladaptive norms fail to prosper in competition with communities whose norms and customs encourage activities that are adaptive for the group (and/or the individuals within it).

The importance of group selection to human evolution remains controversial. But it may well be that adaptive customs in traditional societies depend in part on such selection. We see customs that help preserve expertise because groups that were poor at protecting their cognitive resources have disappeared, along with their members. Likewise groups whose norms encouraged maladaptive ends are no longer around to trouble ethnographers. We see adaptive action because of group-level filtering of communities with maladaptive practices. Even if this mechanism was once important, it can no longer be powerful. The transition to mass society has decreased metapopulation size, and has increased the internal heterogeneity of the remaining populations. These changes depower cultural group selection. So cultural group selection will filter norms and beliefs much less effectively. To sum up the argument so far: heritability declines, and cultural group selection is less powerful. So adaptive

---

[5] Perhaps, for example, there was selectable variation in local practices that support the division of informational labour and specialization – hence accumulating information more efficiently – or practices that support more effective teaching, for example by according high prestige to those prepared to share their expertise.

responses cannot be sustained by population-level mechanisms; lineages are less 'intelligent'. In mass society, adaptive action will depend on mechanisms of individual adaptive plasticity, plasticity of belief and of value.

### 11.4.2 Instrumental rationality in complex societies

Enculturation increases the informational load on adaptive action. Compared with our minimally cultured ancestors, wielding something like Oldowan technology, and living in social worlds roughly comparable to those of chimps in size and complexity, later hominins needed a much richer stock of information. A brief and partial list of the increased requirements might include the following. (i) A much expanded folk psychology: later hominins routinely engage in joint planning and collective action, and this requires information about the capacities and intentions of social partners. (ii) Later hominins need to understand human symbolic systems, most obviously language, but depictive representation, and signals of age, status and role encoded in dress and action. (iii) A successful agent in recent hominin social worlds needs to understand the norms and normatively laden customs of that world. (iv) Our core group size has probably expanded. So we each will need an expanded database of individual agents: a database that will include partial biography and social assessment, not just recognitional capacity (Dunbar 2001, 2003). (v) Human social worlds have become vertically complex. Human social worlds include teams, extended families, clans, hunting alliances and many other stable, functionally important units intermediate between individual agents and the social world as a whole. Effective action requires the capacity to recognize and work within these intermediate units. (vi) Our technology has expanded explosively, and while (especially in contemporary worlds) no one individual is a master of all the technological resources of his or her group, each agent has the capacity to use (and sometimes make or repair) many tools and their products. Compared with our distant ancestors, we are all engineers. (vii) We have long lived in a world of trade and division of labour. So we must know the economic or exchange value of many goods.

Many of these changes have ancient origins, but they have all intensified (or originated) with the transition to large-scale social worlds. Over the last 10,000 years, human social worlds have become larger and more individually heterogenous. They have more hierarchical structure and more occupational complexity. At roughly the same time, human symbolic systems have become more complex with the invention of writing, the elaboration of numeracy supported by increasingly powerful notational systems and with the expansion of depictive representation. Human social worlds have become vertically complex, and more technologically and economically complex. We need more informational resources to make adaptive choices.

Fortunately for our capacity to achieve our goals, our access to information has improved too. Human learning environments are adapted, making trial-and-error learning more effective, and our technology includes informational technology: most obviously language. Moreover, humans invest time in explicit teaching. So despite the expansion of our informational needs, the shift to large-scale societies may not have exacerbated the problem of information stress. The transition to mass society increases informational demands on adaptive choice, but it also provides more information tools. In key domains of action, we have continued access to instrumentally relevant information, and a continued ability to use that information. This explains the viability of rational agent models of human action. In contrast to the models of behavioural ecology, they can be neutral about the connection between utility and fitness.

### 11.4.3 Wanting what our genes need

The course of hominin evolution has transformed not just hominin access to information; it has also transformed the motivational structure of human minds, including making our motivations much more sensitive to our social environment. Our ancestors were not chimps. But I shall follow others in treating chimps and bonobos as rough analogues of the minds and capacities of early hominins. Chimp goals are very different from those of modern humans (Tomasello 2009). Some of the most salient differences are:

(i) Humans are not always good at deferring gratification. But we routinely engage in planned activity whose rewards are hours, days, weeks in the future. Many of us save, deferring reward for years and decades. (We do so, in part, by structuring our environment to make the temptations to take immediate reward less available or less tempting.) Chimps have very high discount rates: they find it very difficult to defer reward.[6]

(ii) Chimps are less strongly motivated by social emotions than are humans. They will engage in some low-cost prosocial helping

---

[6] Though see Rosati et al. 2007; Osvath and Osvath 2008.

(Warneken and Tomasello 2006; Warneken et al. 2007), but there is no sign that they are motivated by such moral emotions as fairness or a preference for egalitarian outcomes. Thus, unlike humans, they seem to act like economic maximizers in ultimatum games (Jensen et al. 2007). In strong contrast to young children, there is no evidence that young chimps find collective activity intrinsically rewarding (Warneken, forthcoming).

(iii) There is no sign that chimp social interaction is regulated by norms that the agents themselves have internalized.

(iv) As with any animal, we feel hungry and thirsty. But the foods we eat, and the circumstances in which we eat, have been much modified by culture, and in ways that vary significantly from culture to culture. Disgusts and taboos show significant cultural variation. In many cultures, resource consumption has acquired an additional signalling function, so our consumptive appetites are mixed with social motivations. Food isn't just fuel; for us, basic biological needs have been infected with cultural significance (Jones 2007). One of the charms of watching chimps at the zoo is their apparent freedom from anything that corresponds to the human emotion of embarrassment, especially in conjunction with lust, hunger or defecation.

(v) We want what others want, because they want it. In part, no doubt, this is a decent but fallible epistemic heuristic: if a book is a bestseller, and I have no reason to regard my own tastes as unusual, I am quite likely to enjoy it too. But as Frank argues (Frank 2000), our interest in what others have is not just instrumental. How I value what I get depends in part on what you get, and on your responses to what I get.

(vi) In contrast to other primates, we have second-order preferences. I can be strongly motivated by the desire for chocolate, but also desire not to want chocolate. While second-order preferences do not inevitably trump first-order ones, and most certainly do not extinguish them, they are not epiphenomenal. In the short-run, second-order desires can block or modify acting on first-order desires, and in the long-run, changes in habitual patterns of action change first-order motivation. I gave up sugar in tea and coffee for health reasons, but no longer have any desire for sweetened caffeine. Mark Hauser discusses more dramatic cases of this interaction, in discussing moral vegetarianism. Such vegetarians often initially find it difficult to give up meat eating, but over time come to find the prospect of meat eating disgusting (Hauser 2006).

As a result of these evolved changes in our motivational structure, our motivations are, at least in part, malleable over an individual's life, as well as on multigeneration scales. To some extent, we learn what we want, not just how to get what we want. Sometimes this is deliberate, as people engineer their first-order desires by changing their way of life to suppress temptation and to enhance the attractiveness of other options. But sometimes there are gradual, unnoticed changes that are side effects of change in the larger environment. Cultural processes install taste and taboo as children mature in their society, but these are not always fixed immutably. The western world has seen major changes in food, dress and sexual norms in the last few decades, and while these changes are reflected most obviously in those who grow up with the new customs, some members of older generations change (perhaps partially) with the times. We are not stuck with the proximate motivators that drove our behaviour at twenty, and not just because of background changes in our intrinsic physiology.

It follows from this plasticity of conscious desire that the targets of individual action are (perhaps profoundly) influenced by variable cultural factors. With the erosion of heritability of phenotypic traits that depend on cultural transmission, and decreased power of cultural group selection, population-level processes are less effective in binding proximal motivation to fitness effects. They are less effective in linking human wants to genetic needs. We can monitor and improve our performance as rational agents in ways that do not depend on population-level processes. But this will not weld human interests to genetic interests. If we get it wrong about what the world is like, often we get a useful though unpleasant signal from the world, and we can use that signal to update and improve our image of the world. An agent can use feedback from the world both to improve his/her stock of instrumental information and to fine-tune techniques for extracting information from the environment. If an agent's proximate motivators come to covary less well with fitness – if the targets at which action is aimed become irrelevant or detrimental to fitness – there need be no consequence in the lifespan of the agent that the agent can recognize and use. Indeed, from the agent's perspective, nothing has gone wrong. A mismatch between what agents value and resources for fitness will not cause the agent's proximate projects to miscarry. So there are no mechanisms operating within the span of individual learning which will systematically cause individuals to unlearn maladaptive values.[7]

### 11.4.4 Upshot

In traditional social worlds, dominated by vertical information flow and group selection, proximate projects guided by maladaptive values will have population-level consequences. Agents with such values will decline in frequency over time. But if vertical information and value flow is less salient and group selection is insignificant, there are no population-level mechanisms which filter maladaptive values (except perhaps in extreme cases). In brief: the transition to mass society need not undermine instrumental rationality; in favourable cases, we can estimate the outcomes of action. But the targets at which we aim need not contribute systematically to fitness maximization. Hence when mechanisms of instrumental assessment are still intact, some form of a utility-maximization model is likely to capture central phenomena of mass societies. In key cases, agents give themselves their best chance of getting what they want; but what they want is no longer reliably a fitness resource. Utility is decoupled from fitness.

### REFERENCES

Binmore, K. (2006). 'Why Do People Cooperate?' *Politics, Philosophy and Economics* 5: 81–96.

Bliege Bird, R., E. A. Smith and D. W. Bird (2001). 'The Hunting Handicap: Costly Signaling in Human Foraging Strategies'. *Behavioral Ecology and Sociobiology* 50(1): 9–19.

Blurton Jones, N. (1997a). 'Bushman Birth Spacing: A Test for Optimal Interbirth Intervals'. *Human Nature*, ed. L. Betzig. Oxford, Oxford University Press, 73–82.

(1997b). 'Too Good To Be True? Is There Really a Trade-Off between Number and Care in Human Reproduction?" *Human Nature*, ed. L. Betzig. Oxford University Press, 83–86.

Blurton Jones, N., K. Hawkes and J. F. O'Connell. (2006). 'The Global Process and Local Ecology: How Should We Explain Differences between the Hadza and the !Kung?' *Cultural Diversity among Twentieth Century Foragers: An African Perspective*, ed. S. Kent. Cambridge University Press, 159–187.

---

[7] Thus in contrast to Don Ross, I do not think that in the long run, fitness must count, and that culturally supported but fitness-eroding values and norms will be filtered out by population-level mechanisms (Ross 2006, 2008).

Boehm, C. (1999). *Hierarchy in the Forest*. Cambridge, MA, Harvard University Press.

Bowles, S. and H. Gintis (2006). 'The Evolutionary Basis of Collective Action'. *The Oxford Handbook of Political Economy*, ed. B. Weingast and D. Wittman. Oxford University Press, 951–970.

Boyd, R., H. Gintis, S. Bowles and P. J. Richerson (2005). 'The Evolution of Altruistic Punishment'. *Moral Sentiments and Material Interests: The Foundations of Cooperation in Economic Life*, ed. H. Gintis, S. Bowles, R. Boyd and E. Fehr. Cambridge, MIT Press, 215–227.

Boyd, R. and P. Richerson (1992). 'Punishment Allows the Evolution of Cooperation (or Anything Else) in Sizable Groups'. *Ethology and Sociobiology* 13: 171–195.

Crosby, A. W. (2004). *Ecological Imperialism: The Biological Expansion of Europe, 900–1900*. Cambridge University Press.

Diamond, J. (1998). *Guns, Germs and Steel: The Fates of Human Societies*. New York, W.W. Norton.

Dunbar, R. (2001). 'Brains on Two Legs: Group Size and the Evolution of Intelligence'. *Tree of Origin*, ed. F. de Waal. Cambridge, MA, Harvard University Press, 173–192.

(2003). 'The Social Brain: Mind, Language and Society in Evolutionary Perspective'. *Annual Review of Anthropology* 32: 163–181.

Fehr, E. and U. Fischbacher (2003). 'The Nature of Human Altruism'. *Nature* 425: 785–791.

Fehr, E. and S. Gächter (2002). 'Altruistic Punishment in Humans'. *Nature* 415(10 January): 137–140.

Frank, R. (1988). *Passion within Reason: The Strategic Role of the Emotions*. New York, W. W. Norton.

(2000). *Luxury Fever: Money and Happiness in an Era of Excess*. Princeton University Press.

Gächter, S. and B. Herrmann (2009). 'Reciprocity, Culture and Human Cooperation: Previous Insights and a New Cross-Cultural Experiment'. *Philosophical Transactions Royal Society B* 364(1518): 791–806.

Gintis, H. (2006). 'Behavioral Ethics Meets Natural Justice'. *Politics, Philosophy and Economics* 5(1): 5–32.

(2007). 'A Framework for the Unification of the Behavioral Sciences'. *Behavioral and Brain Sciences* 30(1): 1–61.

(2009). *The Bounds of Reason: Game Theory and the Unification of the Behavioural Sciences*. Princeton University Press.

Godfrey-Smith, P. (2006). 'The Strategy of Model-Based Science'. *Biology and Philosophy* 21: 725–740.

(2009). 'Models and Fictions in Science'. *Philosophical Studies* 143: 101–116.

Gurven, M. and K. Hill (2009). 'Why Do Men Hunt? A Reevaluation of "Man the Hunter" and the Sexual Division of Labor'. *Current Anthropology* 50(1): 51–74.

Hauser, M. (2006). *Moral Minds: How Nature Designed Our Universal Sense of Right and Wrong*. New York, HarperCollins.

Hawkes, K. and R. Bird (2002). 'Showing Off, Handicap Signaling and the Evolution of Men's Work'. *Evolutionary Anthropology* 11(1): 58–67.

Henrich, J., R. Boyd, S. Bowles, et al. (2005). '"Economic Man" in Cross-Cultural Perspective: Behavioral Experiments in 15 Small-Scale Societies'. *Behavioral and Brain Sciences* 28: 795–855.

Hrdy, S. B. (2009). *Mothers and Others*. Cambridge, MA, Harvard University Press.

Jensen, K., J. Call, and M. Tomasello (2007). 'Chimpanzees Are Rational Maximizers in an Ultimatum Game'. *Science* 318(5 October): 107–109.

Jones, M. (2007). *Feast: Why Humans Share Food*. Oxford University Press.

Jukema, J. and T. Piersma (2006). 'Permanent Female Mimics in a Lekking Shorebird'. *Biology Letters* 2(2): 161–164.

Kaplan, H. (1996). 'A Theory of Fertility and Parental Investment in Traditional and Modern Human Societies'. *American Journal of Physical Anthropology* 101(S23): 91–135.

Keen, I. (2006). 'Constraints on the Development of Enduring Inequalities in Late Holocene Australia'. *Current Anthropology* 47(1): 7–38.

Klein, R. and B. Edgar (2002). *The Dawn of Human Culture*. New York, Wiley.

Krebs, J. and N. Davies, eds. (1997). *Behavioural Ecology: An Evolutionary Approach*. Oxford, Blackwell.

Laland, K. (2007). 'Niche Construction, Human Behavioural Ecology and Evolutionary Psychology'. *Oxford Handbook of Evolutionary Psychology*, ed. R. Dunbar and L. Barrett. Oxford University Press, 35–48.

Laland, K. and G. Brown (2002). *Sense and Nonsense: Evolutionary Perspectives on Human Behaviour*. Oxford University Press.

Marlowe, F. W. (2000). 'Paternal Investment and the Human Mating System'. *Behavioural Processes* 51: 45–61.

Norman, M., J. Finn and T. Tregenza (1999). 'Female Impersonation as an Alternative Reproductive Strategy in Giant Cuttlefish'. *Proceedings of the Royal Society B* 266: 1347–1349.

Osvath, M. and H. Osvath (2008). 'Chimpanzee (*Pan troglodytes*) and Orangutan (*Pongo abelii*) Forethought: Self-Control and Pre-experience in the Face of Future Tool Use'. *Animal Cognition* 11(4): 661–674.

Quinn, T. (2005). *The Behavior and Ecology of Pacific Salmon and Trout*. Seattle, University of Washington Press.

Richerson, P. and R. Boyd (2001). 'The Evolution of Subjective Commitment to Groups: A Tribal Instincts Hypothesis'. *Evolution and the Capacity for Commitment*, ed. R. Nesse. New York, Russell Sage Foundation, 186–220.

Richerson, P., R. Boyd and J. Henrich (2003). 'Cultural Evolution of Human Co-operation'. *Genetic and Cultural Evolution of Cooperation*, ed. P. Hammerstein. Cambridge, MIT Press, 373–404.

Rosati, A., J. Stevens, B. Hare and M. D. Hauser (2007). 'The Evolutionary Origins of Human Patience: Temporal Preferences in Chimpanzees, Bonobos, and Human Adults'. *Current Biology* 17: 1663–1668.

Ross, D. (2006a). 'Evolutionary Game Theory and the Normative Theory of Institutional Design: Binmore and Behavioral Economics'. *Politics, Philosophy and Economics* 5(1): 51–80.

(2006b). 'The Economic and Evolutionary Basis of Selves'. *Cognitive Systems Research* 7: 246–258.

(2008). 'Economics, Cognitive Science and Social Cognition'. *Cognitive Systems Research* 9: 125–135.

Seabright, P. (2006). 'The Evolution of Fairness Norms: an essay on Ken Binmore's Natural Justice'. *Politics, Philosophy and Economics* 5(1): 33–50.

(2010). *The Company of Strangers*. Princeton University Press.

Smith, E. A., K. Hill, F. Marlowe, et al. (2010). 'Wealth Transmission and Inequality among Hunter-Gatherers'. *Current Anthropology* 51(1): 19–34.

Smith, E. A. and B. Winterhalder (2003). 'Human Behavioral Ecology'. *Encyclopedia of Cognitive Science*. L. Nadel. London, Nature Publishing Group. 2: 377–385.

Sober, E. and D. S. Wilson (1998). *Unto Others: The Evolution and Psychology of Unselfish Behavior*. Cambridge, MA, Harvard University Press.

Sterelny, K. (2007). 'Social Intelligence, Human Intelligence and Niche Construction'. *Proceedings of the Royal Society B* 362(1480): 719–730.

(forthcoming). 'Life in Interesting Times: Co-operation and Collective Action in the Holocene'. *Evolution, Cooperation, and Complexity*, ed. B. Calcott, B. Fraser, R. Joyce and K. Sterelny. Cambridge, MA, MIT Press.

Tomasello, M. (2009). *Why We Cooperate*. Cambridge, MA, MIT Press.

Warneken, F. (forthcoming). 'The Origins of Human Cooperation from a Developmental and Comparative Perspective'. *The Evolution of Mind*, ed. G. Hatfield. Philadelphia, University of Pennsylvania Press.

Warneken, F., B. Hare, A. P. Melis, D. Hanus and M. Tomasello (2007). 'Spontaneous Altruism by Chimpanzees and Young Children'. *PLoS Biology* 5(7): e184, doi:10.1371/journal.pbio.0050184.

Warneken, F. and M. Tomasello (2006). 'Altruistic Helping in Human Infants and Young Chimpanzees'. *Science* 311(3 March): 1301–1303.

Weisberg, M. (2007). 'Who Is a Modeler?' *British Journal for the Philosophy of Science* 75(3): 383–404.

Wimsatt, W. C. (2007). *Re-engineering Philosophy for Limited Beings: Piecewise Approximations to Reality*. Cambridge, MA, Harvard University Press.

Winterhalder, B. and E. A. Smith (2000). 'Analyzing Adaptive Strategies: Human Behavioral Ecology at Twenty Five'. *Evolutionary Anthropology* 9(2): 51–72.

Zahavi, A. (1990). 'Arabian Babblers: The Quest for Social Status in a Cooperative Breeder'. *Cooperative Breeding in Birds*, ed. P. Stacey and W. Koenig. Cambridge University Press, 105–133.

Zizsser, H. (1935). *Rats, Lice, and History*. London, George Routledge & Sons.

# Index

Printed in the United States
By Bookmasters